# Praise for *The Urban Far*



*The Urban Farmer* is simply the best guide out there f vegetables for market. Chock full of practical informati , business planning, the best crops to grow, how much land to farm, growing techniques, and how to develop markets, this book covers it all. Curtis Stone shares his hard-won knowledge on setting up and succeeding at small-plot intensive (SPIN) farming in lively, easy-to-grasp prose, in all the detail you'll need to get started. Curtis not only tells us what works, he reveals, based on his own experience, what didn't work for him, and that alone is worth the price of the book. This is a comprehensive real-world manual from someone who's done it, and any market farmer will profit greatly from reading it.

— Toby Hemenway, author of *Gaia's Garden* and *The Permaculture City*

I have no hesitation in saying that *The Urban Farmer* by Curtis Stone is one of the most important, and overdue, books on urban agriculture ever published. It is simultaneously deeply visionary and immensely practical, always a heady brew. It allows us to look at urban land in an entirely different way. If I were 18 again and given this book, it would put fire in my belly and set me on a career path that is cutting edge, deeply entrepreneurial, and profoundly responsible. It deserves to be a best seller.

— Rob Hopkins, Founder of the Transition movement
and author of *The Power of Just Doing Stuff*.

Curtis Stone is at the forefront of a stirring revolution. Urban farming will change what local food means and I know of no other farmer that is as successful at it as he is. And the best part is his willingness to share what is a successful business model. If you're interested in learning to profitably start a farm on a shoestring budget, Curtis Stone is the go-to guy.

— Jean-Martin Fortier, author of *The Market Gardener*

A first-rate, hands-on guide to successful and profitable farming on the very small scale, Curtis Stone's *The Urban Farmer* should be required reading for anyone who thinks that growing food requires hundreds of acres off in the countryside. Highly recommended.

— John Michael Greer, author of *Green Wizardry*

Curtis Stone has artfully blended my three favorite things—entrepreneurship, independence and sustainable food production—into one amazing book. He has also done so in a way that lowers the entry point for anyone who is truly motivated to no longer have any excuse for not getting started. To say I recommend this book highly is a gross understatement. I consider it required reading for anyone with a goal to start a business, not matter what niche they end up in.

— Jack Spirko, TheSurvivalPodcast.com

Urban agriculture is a tradition dating back thousands of years as well as an innovation reshaping modern city design. It's also a lure for a growing number of idealists drawn by a vision of reconnecting with the land while becoming part of the solution. But hold on. Anyone who's tried it as a business knows there's more to urban agriculture than romance. It takes hard work and common sense—two gifts Curtis Stone has in spades, and he's always been generous with sharing it. Local growers have appreciated the lectures and workshops where he spells out the dollars and sense behind growing city food. Now readers everywhere have the opportunity to tap into this valuable resource.

If you're going to invest in your future as city food grower, start with a copy of *The Urban Farmer*.

— David Tracey, author of *Guerrilla Gardening* and *Urban Agriculture*

This book is a treasure for anyone really serious about making a decent living off an urban farm. Back-to-the-lawn urban farming might look easy, but Curtis Stone shows exactly how that "ease" grows out of getting a thousand details right. They're all in this book, generously shared. This is not just a well-written business text, illustrating the myriad technical, entrepreneurial, marketing, accounting, farming and people skills Curtis developed to work smarter, not harder. It is even more a quintessential "how-to" manual, taking the reader step by step by step to the roots of running a profitable urban farm.

— Peter Ladner, author of *The Urban Food Revolution*
and a long-time urban food gardener

# THE URBAN FARMER

# THE URBAN FARMER

## GROWING FOOD FOR PROFIT ON LEASED AND BORROWED LAND

CURTIS STONE

Cover design by Diane McIntosh.
Upper left cover photo © Andrew T. Barton.

Unless otherwise noted, all cover and interior photographs by Curtis Stone.
All interior illustrations by Anthony Ross (flexanimousart.blogspot.ca).

Printed in Canada. Fourth printing May 2017.

Funded by the Government of Canada | Financé par le gouvernement du Canada | Canada

Inquiries regarding requests to reprint all or part of *The Urban Farmer* should be addressed to New Society Publishers at the address below.
To order directly from the publishers, please call toll-free (North America) 1-800-567-6772, or order online at www.newsociety.com

Any other inquiries can be directed by mail to:

New Society Publishers
P.O. Box 189, Gabriola Island, BC V0R 1X0, Canada
(250) 247-9737

LIBRARY AND ARCHIVES CANADA CATALOGUING IN PUBLICATION

Stone, Curtis, 1979-, author
The urban farmer : growing food for profit on leased and borrowed land / Curtis Stone.

Includes bibliographical references and index.
Issued in print and electronic formats.
ISBN 978-0-86571-801-2 (paperback). — ISBN 978-1-55092-601-9 (ebook)

I. Title.

S494.5.U72S86 2016 630 C2015-906769-3
C2015-906770-7

New Society Publishers' mission is to publish books that contribute in fundamental ways to building an ecologically sustainable and just society, and to do so with the least possible impact on the environment, in a manner that models this vision. We are committed to doing this not just through education, but through action. The interior pages of our bound books are printed on Forest Stewardship Council®-registered acid-free paper that is 100% post-consumer recycled (100% old growth forest-free), processed chlorine-free, and printed with vegetable-based, low-VOC inks, with covers produced using FSC®-registered stock. New Society also works to reduce its carbon footprint, and purchases carbon offsets based on an annual audit to ensure a carbon neutral footprint. For further information, or to browse our full list of books and purchase securely, visit our website at: www.newsociety.com

# Contents

Foreword, by Diego Footer . . . xi
Preface . . . xv

1 A Farm in the City . . . 1
1. Why Urban Farming? . . . 3
2. Connecting the Dots: An Urban Farmer's Place in the Community . . . 7
3. Quick Breakdown of Economics . . . 11

2 A Viable Farming Business On ½ Acre Or Less . . . 13
4. The Zones of Your Farm and Your Life . . . 15
5. Crops Better Suited for the City . . . 19
6. Introduction to Urban Infrastructure . . . 27
7. Start-Up Farm Models . . . 29

3 The Business of Urban Farming . . . 37
8. Starting Small . . . 39
9. Market Streams . . . 41
10. Working with Chefs . . . 51
11. Labor . . . 55
12. Software and Organization . . . 61
13. Self-Promotion . . . 73
14. Finance Options . . . 75

4 Finding the Right Site . . . 79
15. Scouting for Land . . . 81
16. Urban, Suburban and Peri-Urban Land . . . 89
17. Multiple or Single-Plot Farming . . . 93
18. Urban Soil . . . 95
19. Land Agreements and Leases . . . 99
20. Urban Pests . . . 101

5 Building Your Farm, One Site at a Time . . . 105
21. Turning a Lawn Into a Farm Plot . . . 107
22. Choosing A Site . . . 115
23. Garden Layout . . . 119
24. The Perimeter . . . 121
25. Irrigation . . . 123

6 Infrastructure and Equipment . . . 133
26. Base of Operations . . . 135
27. Tools . . . 143
28. Special Growing Areas . . . 147
29. Inexpensive Season Extension . . . 151
30. Transportation . . . 155

7 Operations . . . 157
31. Work Smarter not Harder . . . 159
32. Harvesting . . . 167
33. Post-Harvest Processing . . . 173
34. Portioning and Packing . . . 179

8 Production Systems . . . 185
35. Beds for Production . . . 187

36. Planting . . . 195
37. Microgreens . . . 199
38. Extending the Season . . . 205

9 Basic Crop Planning . . . 211
39. Determine Your Outcome . . . 213
40. The Base Plan . . . 217

10 Crops for the Urban Farmer . . . 223
41. Parting Words . . . 249

Acknowledgments . . . 251
Glossary . . . 253
Endnotes . . . 255
Index . . . 257
About the Author . . . 265

# Foreword

*by Diego Footer*

Do you dream about becoming a farmer... making a living with your hands in the soil, being outside far away from the confines of the cubicle, working your own piece of land and growing the nutrient-dense food that you want to eat?

It's a nice thought. A worthy ambition.

But if you are seriously going to go down that road then you need to ask yourself: *How are you ever going to make a living farming?*

The concern around farming being a viable and profitable career is legitimate. And the talk of farming NOT being profitable is more truth than exaggeration. Most farmers struggle earning a living farming; recent USDA data supports this:

> Given the broad USDA definition of a farm, many farms are not profitable even in the best farm income years. The projected median farm income of −$1,558 is essentially unchanged from the 2014 forecast of −$1,570. Most farm households earn all of their income from off-farm sources—median off-farm income is projected to increase 4 percent in 2015.[1]

If you have to work another job in order to be a farmer, it isn't good, and you probably won't be a farmer for very long.

Over the past few years with Permaculture Voices, I have spoken to numerous visionaries such as Joel Salatin, Michael Pollan, Geoff Lawton, Mark Shepard, and Allan Savory about the future of agriculture. I understand what's possible and what could be, and I also understand that things need to change.

During that time I have also spoken to countless people looking to break into the farming business and make agriculture their future. These people dream about being farmers—the next wave of farmers, the 20-and 30-somethings that aspire to follow in the footsteps of people like Joel Salatin and break into farming. Through these conversations I have come to fully understand the dream and the lure of the farming lifestyle, but also the obstacles that hinder making that dream a reality.

These obstacles include:

- The high cost and limited access to land
- High capital costs for equipment
- The lack of an efficient distribution system for small farmers
- A broken food system that values cheap food over high-quality, nutrient-dense, locally raised food

For many people, these obstacles are too much to overcome. Costs are too high, prices are too low and margins are too thin. As a result, the dream of farming fizzles out.

Profitable farming is possible, but farming profitably may look different in reality than the idea of farming that you have in your mind. Many people are lured into farming by icons like Joel Salatin who are farming on a lot of acreage. So naturally many people think that in order to be profitable they need to farm a lot of land.

Like many of you, I thought farming had to be big. Then I met Curtis, and he shattered that myth.

I first met Curtis through my work with Permaculture Voices. When I learned about what he was doing, I was a bit taken aback. Here was this guy farming in the city and doing everything by bike. This sounded ideologically great, but how serious an operation could it be? After all he was doing it by bike in the city. Legit farming has to be big right? Wrong. After diving a little deeper I was floored. Here was a guy farming in the city by bike and absolutely crushing it. He was making a consistent five-figure income farming, more than most farmers than you will ever talk to, and he was doing it on one third of an acre. One third of an acre, that's it. It was impressive. Curtis's farm was simple, efficient and effective. It wasn't high tech. This wasn't hydroponics or aquaponics. This was old-fashioned, in-the-ground farming. It didn't seem possible, but it was.

What Curtis is doing is special. It is very different from what you would expect when it comes to the traditional farming model. It is small; it is in the city; it is low tech; it is being done on land that he doesn't own; and it is profitable. It's a paradigm shift.

I have stood on Curtis's plots in Kelowna, British Columbia, and seen his systems at work firsthand. They are the real deal: small-scale, bio-intensive production that is efficiently and effectively managed in a way that is profitable. Curtis's systems work, and this book explains those systems in great detail.

I want to emphasize that these systems have been field tested over time. That's important because there are a lot of theorists out there with great ideas but little actual experience or on-the-ground projects to back themselves up. Curtis backs it up. This is not a book about untested theory; this is a book based on real-world experiences. Curtis has been a profitable farmer for six years. That is his job. His only job. He is supporting himself and earning a living wage farming—as previously stated, this is a rarity in farming.

Why has Curtis been so successful—and how can that help you as a new farmer?

I think it boils down to a few things:

- He markets and brands his farm and his products very well.
- He is extremely efficient and effective with everything that he does on the farm.
- He approaches farming as a business with strict adherence to the bottom line.
- He takes meticulous records, analyzes them, and implements changes based on that analysis.
- He has the ability to solve problems on the fly and quickly adapt.

Most farmers don't do all five of these things well. In fact, many farmers don't do any of these five things. And as a result they don't succeed and stay farmers for very long.

What you will notice about these keys to success is that they aren't directly farming tactics or techniques. They are much more macro, and they are related to the specific methodology and mindset with which Curtis approaches farming. It is this methodology and mindset that has helped Curtis to be successful and profitable as an urban farmer, and it is this approach to farming that I believe will have the greatest effect on your success.

This approach to farming has really resonated with me. As an engineer, I realize the importance of systems thinking and how important it is to dissect larger problems into solvable smaller problems in order to move forward. This is essentially what Curtis does, has done and how he has arrived at his current method of farm production. He has paid special attention to what has worked and what hasn't and then focused on what has worked, dropping off what hasn't. Pay attention to how Curtis approaches farming as you go through *The Urban Farmer* and think about how you can apply his methods to your situation. But don't overthink it. There is no need to reinvent the wheel or overcomplicate things; what Curtis does works. Adapt his techniques to your situation and your market, and do it.

Success in farming is possible. But in order to be successful as a modern-day farmer you need to be more like a modern-day tech entrepreneur than a peasant agrarian of yesteryear. The image that you have in your head of a farmer and the farming lifestyle isn't the image of reality that is going to get you where you want to go. The traditional farming model is broken and it doesn't work.

For the next generation of farmers I think that Curtis brings a different model to the table, and at the end of the day it's all business when it comes to running his farm. The lifestyle is there, but the business comes first.

A lot of new farmers struggle because they follow their hearts instead of approaching farming as a business. As a result they end up letting their hearts down and end up out of farming.

The common causes of failure are widespread: growing crops without a specific market in mind, taking on too much land, not planting densely enough, approaching farming as a lifestyle and a hobby—not a business. All of these causes of failure can be addressed if you approach farming strategically as Curtis has done.

If you want to take on farming as a career then I believe that the information in *The Urban Farmer* can drastically help you skip what doesn't work and focus on what does. I believe that if you model your farm's systems after Curtis's then you can be successful and profitable as a farmer. It won't be easy, and it will take time. You have to be willing to put in the hard work, learn from your own experiences and push through hard times. When other people would quit, and when you want to quit, you need to push on and adapt. The obstacles preventing you from getting into farming can be overcome.

Profitable farming is possible.
The methods are in this book.
The rest is up to you.

— Diego Footer, Founder
Permaculture Voices

# Preface

## Backstory

My entire life I wanted to be a rock 'n roll musician; even from the time I was 16 years old, I was playing in bands with other young men in my town. We would go on weekend tours to nearby cities like Vancouver and Seattle. Playing music was the only thing I actually enjoyed doing, and it was all that I wanted to do. After finishing music school in 2002, I formed a group called People for Audio, and we moved to Guelph, Ontario, to spend a year writing music, recording, and gigging in and around southern Ontario. During this period, and for nine years after, I would travel back to BC for two to three months in the spring to early summer and work as a treeplanter. It was something besides music that I was fairly good at, and it allowed me to make enough money, that I could afford to work only part-time, or sometimes not at all, during the winters so I could focus more on playing music. Perhaps a seed was planted in me at this time: I discovered that I loved working outside and being engaged in nature. In 2003, I and the other main members of People for Audio moved to Montreal together, and that's where I stayed for another six years. The band did a lot of recording and touring in that time, and for most of those years, playing in a band was all I thought I was meant to do.

Growing up as a bit of a "rebel without a cause," I got into punk rock music at a pretty young age, and through that I discovered thinkers such as Noam Chomsky, which led me to follow geopolitical issues with great interest and despair. Back in my punk rock days, I heard the old cliché, "If you're not part of the solution, you're part of the problem," and it has stuck with me ever since. It was around late 2007 when I really started to be personally bothered by a lot of what was happening in the Western world and abroad: endless wars, environmental degradation and an economic system that benefited only a small number of elites. I sometimes went to bed very restless because of all the injustice I saw in the world. My music career at this time was starting to slow down; the band was in the process of breaking up, and this forced me to reevaluate my life in a way that I had never done before. Everything up to that point had led me to believe that my one true calling was to be a musician, but it was obviously not working very well for me.

The winter of that year, I was working at a screen printing shop during the day and spending a lot of my nights cruising the Internet, learning about living off the land, natural building and organic agriculture. I became obsessed with looking for alternative ways to live on this planet in a sustainable way. I felt like I had to take control of my life so that I could simply live by my values. So many things that I saw in the world disgusted me. The way in which we who live in affluent countries travel, eat and make money all made me sick. It was like everything I saw myself do in my day-to-day life was detrimental to the environment in some way, and this caused me to fall into a depression unlike any I had ever experienced. I guess you could say it was a combination of realizing that music might not have been the right thing for me to do in life as well as realizing that I was, in no way, contributing to solutions. I felt like more part of the problem than ever. Cruising around the web one night, I discovered a program called WWOOF (World Wide Opportunities on Organic Farms),[1] and from there I felt like there was an answer for me. This led me further down a path to organic farming, living off the grid and alternative energy. At this point, I started to realize that I needed to get out of Montreal and begin working toward something. I didn't know exactly what that was yet, but I started to put one foot in front of the other, and I drafted up a plan for where I wanted to be in five years. I decided that I needed to live off the land in some way, and so I planned to work the next five years in BC as a treeplanter, working really long seasons, where I could save up a lot of money each year, with the goal that in the end I'd have enough money to purchase a piece of land and begin my off-grid homestead. The plan also included doing a WWOOFing trip for the first part so that I could learn some basic farming skills. I planned to do a trip by motorcycle through BC and the southern west coast of North America, where I would visit farms and homesteads.

In late March of 2008, I left Montreal and headed back to BC, with the plan of treeplanting for the spring and summer, then embarking on this motorcycle trip along the west coast. When I arrived in BC, I began working with a new company and immediately made friends with a fellow named Jason who had a lot of the same ambitions as I had. We became pretty close quickly, and we would often spend the drive to work talking about living off the land and how we both had plans to do so eventually. He told me about a trip that he had taken a year earlier, when he rode his bicycle across the USA. The stories were unbelievable, and I was so inspired by them, that I decided to change my plans of motorcycling down the coast, and ride a bicycle instead. The rest of that planting season, I spent my nights off planning this trip. I was going to ride from Kelowna all the way down the coast!

On August 18, 2008, I left on a trip that would change my life forever. I met incredible people who were generous and kind, and it seemed like there was just one serendipitous experience after the next every place I went. I visited many farms, off-grid homesteads, ecovillages and people just living by their values. What I learned on this trip wasn't so much about how to live off the land, but more about myself: If you wear your values on your sleeve (which is what I was doing in a way), people will approach you all the time. Especially the kind of people you want to be approached by. So, not only was I inspired by all the amazing individuals I met on the trip, but they were also very inspired by me. These interactions totally reshaped my perception of the world, because I realized that I could have a profound effect simply living by my values and demonstrating them in some way. By the time I got to San Diego, I already felt like I could do anything, though my original plan had been to go a lot further. Riding a bike a hundred miles a day every day not only gets you to an amazing physical state but also a mental one. By traveling alone in such a way, I was forced to reach out to people, and I learned that showing a bit of vulnerability can open you up to people in a way I had never experienced. I guess I lived my life before with such a hard sense of pride and arrogance that it often turned people away from me. But this experience taught me it was better to be welcoming and open.

## Why I Wanted to Farm

After returning from my bike tour in November of 2008, I spent that winter reading more books on farming and did a lot of research on how I could do this. I knew then that I wanted to farm somehow, but I still wasn't totally sure how that was going to look. The biggest problem I kept coming back to was that land in BC was so expensive. This was going on year two of my original five-year plan, and I was on my way to making things happen. I had less money saved then I'd hoped, but I had done a lot of research on farming and felt like I knew a thing or two. I had read all of Eliot Coleman's books as well as John Jeavons'. So by now, I knew some kind of small-scale intensive farming was the thing for me. But still the problem of land access kept coming up. I had a friend visit me that winter; we talked about farming, and I explained to him about how buying land was such a challenge. He told me he had heard about a thing called SPIN farming, but he didn't really know that much about it. He mentioned that a farmer using these methods could make $100,000 on an acre of land. When he told me that, my arrogant side immediately dismissed it. From what I read of Eliot Coleman's work, $20,000 per acre was a very high standard for intensive farming. I thought my friend was crazy. Later that spring, I went back to the forests of coastal and interior BC to do another season of treeplanting.

This time around, I began to get a lot more burnt out with this work. At this point it had been my ninth year of planting, and my body was starting to tell me that it was enough. Having come back from that bike trip feeling so energized, I felt myself sinking back into a depressive spiral as I started to feel discouraged that I might not be able to save enough money in time for my five-year goal: to buy land. Also, I knew I had to quit planting, and the dream of buying land to homestead was fading away. About three quarters of the way through that season, on one of my days off I went to visit a friend who was also treeplanting in a nearby town. We got into how discouraged I was about not being able to afford land to farm on, and how I couldn't see myself making it through another planting season. He also mentioned, as my other friend had earlier, SPIN farming. At this point, since two of my good friends mentioned it, I decided that I might as well look into it further.

That planting season wrapped up in June, and when I got back to Kelowna I began reading about this SPIN farming thing. I couldn't believe what they were claiming, so I searched around for anyone that was actually doing this in BC. I found a guy my age named Paul, who was running a SPIN farm using his bike in Nelson, BC (which is a pretty incredible feat, in that Nelson is a town that is as hilly as San Francisco). I contacted Paul, and he was so generous with his time. He allowed me to interview him for a couple hours, and I made tons of notes about his experiences. From that point, I knew that I wanted to do in Kelowna what he had done in Nelson, and I wanted to do it pedal powered just like him as well. I was totally inspired, and I knew that this is exactly what I wanted to do.

### My Barriers Were My Solutions

Having spent the later part of that summer reading more about SPIN farming, as well as going online and searching for more examples of people doing urban farming, I was certain that this is what I was going to do. I started telling all my friends and family about it, and it didn't take long until I had secured one piece of land. It was an urban lot in the downtown of Kelowna that belonged to the family of a longtime friend of mine. It was a double lot, altogether nearly ½ acre with a 2,000-square-foot heritage home on the one lot with a front and backyard, each 2,400 square feet. Next to it there had been a home that had burned down a few years before. Those property owners had torn the home down and completely removed it from the land. So, all that was remaining there was a big hole where the foundation once was. The family was very kind to offer me the property; in exchange I would look after the entire property and provide them with a basket of veggies each week once I got into production. A group of people was renting the house at the time, and the owners were spending a fair amount of money each month to keep up the landscaping, which had fallen into disarray. It seemed like I showed up at the right place at the right time. The owners needed someone to manage the property, and I needed a place to farm: it was win-win on both counts. In August 2009 I started to develop the property. We started by building a fence around the perimeter, as it was on a corner and very open, so fencing it was a must. We also had to bring in all new soil on the side lot, as the existing soil was basically just builders' rubble. The landowners were very generous in that they paid me and a friend to build the fence, and they covered the cost to bring in the new soil. All in all, it cost around $8,000 to build the fence and bring in the new soil. By early October that year everything was set for the next season. Once we had the grass stripped off, I formed out beds and planted the area with a fall rye cover crop. Not much else happened on the property until next spring.

Later that summer, I wanted to do something with farming, to get as much hands-on experience as I could before the winter, but there wasn't much planting I could do that late in the season. I had heard of a group in Victoria, BC, called Pedal to Petal, which ran a pedal-powered compost pickup service; I was inspired by what they were doing, so I figured I'd spend the fall and winter doing something similar so that I could at least learn a thing or two about compost. I went to an ice cream parlor in town and asked if I could take all their scrap buckets; they were pleased to give them to me, as they would otherwise have been garbage. I gave one bucket to each of my friends and said, "Save me your vegetable scraps, and I'll pick them up once a week with my bike." It didn't take long until this little composting program took off. I was spending around 20 hours a week picking up, turning piles and finding dry brown material for the piles. Also, a restaurant that a friend worked at was interested in having me take their scraps as well. It was sort of a foot in the door for getting to know a chef, and it led to many more of those kind of relationships.

Spending time each week for the fall and winter with the compost program kept me pretty busy, along with reading books about farming and gardening. It didn't take long for people in the neighborhood to see that something was going on at this site, and before I knew it I had lots of people coming around to ask questions. Then it was local newspapers and radio stations. By October of that year, still not even technically farming, I was being asked by garden clubs and schools to come and speak about what I was doing. I felt kind of weird about it, because I still didn't really know anything on the practical side, but I guess I did in theory and I was a constant state of learning. So I became reasonably articulate about urban farming and what I was doing, mainly because I had to explain the same thing so many times over. By about late winter that year, I started to prepare my long season crops such as tomatoes, peppers, onions and such. Because of all the press I had done over the course of six months, a lot of people in town had come to know me as the urban farmer guy, or the compost kid.

By this point, I remember experiencing phases of anxiety because I actually didn't know what I was doing at all. I remember waking up some nights in a cold sweat thinking, "Holy shit, am I actually going to do this? All these people in town think I'm this great urban farmer, but I've actually never fully grown a vegetable in my life. What if I fail? How embarrassing would that be?" I basically learned to ignore these negative thoughts and constantly kept a positive mind set. I really followed the saying "Fake it til you make it;" that became my mantra. I was, however, working very hard to consistently learn new things, and I spent a lot of time seeking out mentors in the community for advice. If it weren't for some of the elder organic farmers and gardeners in my area, things may have gone differently for me. I found so much value in speaking with growers who did things completely differently in a production sense, than I did but whose knowledge of plants, pest cycles, soil fertility and even life were so paramount to my success early on. I continually tried to listen, learn and not be afraid to show vulnerability and ask questions that I thought were stupid. I always asked questions, and I never pretended that I had all the answers. I still hold those values to this day. These were things that I learned on that bike tour down the west coast that have been invaluable to my success in farming and in life.

## Lessons from My First Four Years

In my first season of farming, I learned a valuable lesson within the first three months, and I want to instill it into you, the reader. Don't take on too much right away! Start with ¼ acre or less! I started with ½ acre. I actually started earlier that year having just one plot of 6,000 square feet, and I should have kept it at that. One of the challenges for an urban farmer is the fact that so many people love the idea that, once they see you in motion, the land offers flood in. That's exactly what happened to me. Between November and March of 2010, there were so many articles in the local papers about what I was doing that I literally got one phone call a day with land offers. It was absolutely ridiculous, and it was so hard to not be so excited about it that I ended up taking on way too much. During that season, I was farming on seven plots of land totaling ½ acre. My main problem in that year was that I didn't discriminate on location as much as I should have, and because the farm at this point was totally pedal powered, I wasted a lot of my time biking from plot to plot. I also grew far too many types of crops, and most of them barely made much profit. For the most part, this season was still successful, looking back. I grossed $22,000 from being in production from May 15 through October 31. I had one helper working nearly full time for that season as well. This was definitely my hardest year, and I worked nearly 100 hours every week for the entire year, even after the main season was done.

For the next two years my total land mass stayed around the same (at ½ acre), and I had a full-time helper as well. The gross profits of the farm grew without growing the land mass. In 2011, I did $55,000 gross on ½ acre, and in 2012 I did $78,000 on less than ½ acre. From this point I started to learn that it wasn't so much about how much land I was farming but the crops that I grew and the markets I pursued. As a trend over those first three years, I initially sold mostly to farmers markets, but less and less as the years went on. I found that spreading over a few market streams—such as Community Supported Agriculture (CSAs), farmers markets and restaurants—allowed me to move a lot more product, because what didn't sell through one market stream I could sell through another. The main market stream that took off in my third year was restaurants. Having been introduced to one particular chef named Bernard Cassavant, my business almost doubled over the course of a month. Bernard was a high-profile chef in BC, and when he started to buy my stuff, more and more restaurants followed suit. It was an exciting time as his restaurant alone would sometimes have orders nearing $1,000 a week.

During 2013, I took on a partnership with a friend and merged our farms. He had been farming for one year. Together we farmed 2.5 acres and grew around 90 types of vegetables. We had one site (that was previously his) at 2 acres, and the other ½ acre consisted of all my urban plots, with one peri-urban plot across the street from his two-acre site. This was definitely the biggest year as far as sales went. We grossed around $130,000, but our expenses grew so high that it was hard to make a profit in the end. At the height of the season, we had around eight people working nearly full time, and the farm became very top heavy: too much management, too many crops and too much land. Our partnership ended after that season, and we went our separate ways. In the end, it worked out better for both of us. I reiterate the lesson I should have really learned after my first year: Don't take on too much! Start small, and grow slowly!

### Green City Acres: A Commercial Farm Grossing $75K on 15,000 Square Feet

In 2014, I drastically reduced the size of the farm down to one third of an acre on five different plots that were mostly centralized within one third of a square mile. We focused on growing around 15 different crops, which I have discovered over the years are the most lucrative, based on price and days to maturity. We serviced seven restaurants, two wholesale delivery services and one weekly farmers market.

On average I worked 40 hours a week and enjoyed the most laid-back lifestyle I had ever experienced in my five years of farming. In the summer, when production was steady and we were past the set-up phase for the season, we actually worked less. Monday through Thursday, we were usually done by 2 PM and got to spend the rest of the day at the beach relaxing. I had only one part-time employee, who worked 16 hours a week, and I had a few neighbors who helped around the farm. This help was all offered in trade for vegetables: a simple and mutually beneficial trade which made everyone happy.

The main thing that changed from 2013 to 2014 was the dropping most of the crops and land. During the winter of 2013, I tried to figure out why we had made so little profit compared with previous years. I dumped all the profits we made from each crop into a spreadsheet and sorted it according to their total sales. What I discovered was that about ten crops made almost 80% of the income of the total farm, and all of those crops were the ones grown on the small lots in the downtown core. I also discovered that our CSA program, which was the largest I had done in years, had a fairly low return, based on the labor and time it took to maintain it. Going forward into 2014, I decided to cut the CSA and cut about 80% of my crop production; I decided to specialize in 15 crops, focus on my restaurant clients and keep my weekly farmers market. It was a pretty huge shift, but 2014 turned out to be my best season ever. I made a much higher gross and worked far fewer hours.

This is what I want to demonstrate to you in this book: a better way to farm, where you can achieve a lifestyle that is personally sustainable and economically profitable.

PART 1

# A FARM IN THE CITY

# Why Urban Farming?

You've probably heard the term the *end of suburbia* before. In fact, a very well-known film was actually made about the whole concept. The basic premise is that as fuel prices increase, living in the suburbs will become less economically feasible for average North Americans; the cost and time it takes to drive into the city for work will outweigh the benefits of living in the suburbs, and this will cause their imminent collapse. Hence the term, the *end of suburbia.* You can look at that in one of two ways:

1. The decline of real estate values and mass exodus from the suburbs will turn them into ghost towns.
2. There is a huge opportunity to repurpose these places into modern day, self-reliant farming communities.

This book will show you how option #2 is possible.

Let's look at some facts. Right now in the US, there are 40 million acres of lawn. Between 30% and 60% of the fresh water in cities is used to water those lawns, and 580 million gallons of gasoline are used to mow them.[1] When we factor in all the costs it takes to maintain a lawn—such as watering, mowing, weeding and manicuring—it's easy to come to the conclusion that a lawn is nothing but a cost center, one that a lot of North Americans simply cannot afford.

But what if we changed our thinking about lawns? We can tackle two huge problems:

1. Lawns are unsustainable in many ways
2. Access to land is a major barrier for most young people who want to enter the agricultural sector

and create one great solution. Lawns, particularly in suburbs, offer great opportunity for new farmers because:

1. Land is abundant. The average home in the US has an average of .2 acres of land. That's around 8,000 square feet.[2]

2. Using land without owning it removes the idea that one must own land in order to be a farmer.
3. All of that land sitting in lawns now becomes a great place to farm.

What if we could repurpose the suburbs to be the new frontier of localization? What if all of these suburban streets turned into areas for transition, reeducation and abundance? I believe this is not only a possibility but an inevitability.

There are a number of reasons why farming in the city is more profitable, but there are also a variety of reasons that make it very advantageous: access to markets, low start-up and overhead costs, better growing conditions with warmer climates and easy access to water.

### Advantages of Being Urban:

#### *Market Access*

*Market access* has to be the single greatest advantage that benefits urban farmers. When you live and work in the city, you live and work in the market that you're supplying. You don't have to travel very far to sell your product, and for the most part, your product will sell itself. When I deliver to restaurants in the downtown core, I am a five-minute bike ride away from them. Not only is that a talking point that those chefs will boast about to their customers, but it is also a huge advantage to me as far as saving time and energy in transport.

Delivering product that was harvested just blocks from where it is consumed has huge marketing appeal. Our farmers market is a five-minute drive or ten-minute bike ride from our base of operations. One advantage to this, besides bragging rights, is that, if I sell out of one particular item during the market day, I can buzz home quickly on my bike and get more. I call this topping up, and I've done it many times. From my proximity to the market, in 30 minutes, I can ride home, harvest some greens, bring them back to the market and bag them up there. What other farmer has the ability to do that?

#### *Low Start-Up and Overhead*

Farming in city greatly reduces the barriers to entry because you no longer need to think about buying land: it's available everywhere. If you can make enough income on small lots, you don't need the heavy machinery and infrastructure that is required for farming in the traditional sense. Infrastructure is simple, small and cheap.

#### *Better Growing Conditions*

A city is always a few degrees warmer than the countryside. This is called the *heat island effect.* With concrete and buildings everywhere, the city will absorb heat during the day into all that thermal mass, and the heat will release during the evening. This is very noticeable during the summer: if you're riding a bike or driving your car past an open field, you'll immediately notice a drop in temperature. It's because you're leaving the thermal mass of the city that you feel that heat drop. In the downtown of my city, I'm in climate zone 6b, and people just a mile and a quarter out are around a climate zone 5. That's huge difference in frost-free days. In fact, I have at least 30 more frost-free days downtown than farms everywhere else in my area.

The other growing advantage is *microclimates*. When farm plots are surrounded by buildings, walls and fences, these can protect your crops from severe wind. Also, each plot will have its own set of unique characteristics, making some plots better for certain crops than others. This urban climate offers a nice diversity of growing conditions for your farm.

#### *Pests and Weeds*

Pest problems do exist in the city, but when you're farming on multiple plots, they can easily be avoided by simply running away. If flea beetle becomes a problem at one site, you can stop planting that crop there, and start it somewhere else. The pest can't follow the crop around in the city because of its many obstructions and barriers. And weed problems just don't exist in the way they do on rural farms. With barriers and obstructions, there are far fewer weed seeds blowing in from all angles. In the past, I farmed on two peri-urban plots where the neighboring properties were just open fields. The weed problems on these plots were day and night compared to the urban plots.

#### *Access to Water*

Accessible water offers a huge advantage when you compare rural to urban farms. So many farms in the countryside have to wait for water from a river during the spring, and that can delay crop production. Also, well water can become contaminated by neighboring farms or industries. Accessing water on urban lots is in most cases as simple as connecting to the faucet on a house. The water is clean and has ample pressure.

### The Social Connection

Over the years, I have met a large number of my customers, simply because they walked by my farm plots. Every neighborhood where I have a plot, I have a different set of neighbors, just as if I were living on that street. By working in these garden plots over the years, I eventually get to know most of the people on that particular street. This is a very nice thing, as I end up making a bunch more friends.

I can't tell you how many times on a weekly basis I'm visited at my market booth by neighbors from my farm plots. Not only do they become shoppers, but they end up bringing their friends too, who also become customers. There's an old saying "A satisfied customer is your best salesperson." For the urban farmer, it's more like "Your neighbors are your best salespeople." One of the best advantages to having multiple locations is that you just have more potential to build social capital in more neighborhoods.

*Social capital* describes the relationships you build with people over time as a form of capital. Having good relationships with people can turn into many opportunities like favors, connections and influence. These rewards are like money you can save in a bank, except you don't lose any economic value to income tax, sales tax or inflation. No government official can steal social capital from you. It's what you build with people by just making friends, sharing information and feeding the community!

# Connecting the Dots: An Urban Farmer's Place in the Community

### More Than Just a Farmer

What makes urban farming so different than other forms of farming is that you are working in areas where there are many more people, and because of this you will meet more people. An urban farmer has the ability to be more than just a farmer. You can be an *educator* simply by showing people, on the street level, how to grow food. You will connect more people to the idea of *localization* simply by demonstrating it on a day-to-day basis. And you may even get to act as a broker for rural farmers, bringing some of their products that you don't want to grow into the city, helping them access some of the connections you have there.

### The Frontier of Localization

As an urban farmer, you are at the forefront of the local food movement not only because you are growing food to feed locals but because you are demonstrating urban farming to them and leading the way. This has a considerable ripple effect in all the neighborhoods that you will farm in. I have seen these effects firsthand, and it never ceases to amaze me how powerful this can be. In every place that I have had a farm plot over the years, I have seen at least ten people in all those areas start to garden passionately. Whether I was the sole reason they did so is hard to say, but what I can say with certainty is that, since I have been farming in their neighborhood, people have learned a lot about how to grow intensively, and this has been a major motivator to getting them to either start or increase the scale of their gardens. I know this because I hear about it on a day-to-day basis.

### Education

By default, many urban farmers end up becoming educators. Just by farming someone's front yard, you will in many, many instances end up talking to the neighbors about what you're doing. And, the longer you stay there, over the years you're going to be known in that neighborhood for providing a lot of valuable information about how to garden. Before you know it, people will be waiting for you to arrive at your farm plot so they can ask you questions about their gardens. After some time doing this, you're going to get good at it, because you're going to be articulating the same points so many times over that you're going to start to sound like an educator.

This is exactly what happened to me. I had no original intention of being a public speaker, educator or writer. All I wanted to be was a farmer so that I could live by my values. People started to talk to me at my farm plots on an ongoing basis. They'd ask me questions about what I was planting and when to plant it. The more experienced gardeners in the area were always amazed at how early I planted things out into the ground. At first, I was getting many comments like, "Oh, I think you're planting a little too early here; aren't you worried about the frost?" And then after a while, when they saw how I was doing it, they stopped pretending to know more than I did and just started asking questions to help them in their gardens. The first speaking gigs I ever did were at garden clubs, addressing a bunch of retirees/master gardeners. At first I felt ridiculous telling people who had been gardening for 40 years about urban farming, but then I realized that there was something still so new about what I was doing that it could be valuable to them. For people like these, it wasn't so much about the details of how I did this or that but more about why I did it and the fact that I was doing it. I was making a living from essentially gardening in people's yards. You must start with the why. The *what* is important, but the *why* is what pulls people in.

The garden club invitations led to invitations from high schools, then colleges, then universities and conferences. And in less than a full year from when I started, people were actually paying me to come and teach them about urban farming. Granted, there were parts of my past that lent themselves to my being a public speaker: I had been a preforming musician for 16 years of my life, and I realized that what worked for me might not work for everyone. The point here is that, because of the public nature of urban farming, you're going to be communicating what you're doing with people on an ongoing basis. And because of this, I believe that urban farmers have a responsibility to bring their message to the people. No matter how you look at it, you will, whether that is your intention or not.

### The Rural to Urban Connection

I believe there are many opportunities for farmers to work together, and one I see becoming much more important, as urban farming becomes more widely adopted, is connecting rural and urban farmers. Let's face it, some crops just don't make sense to grow on an urban farm. Anything that has a really long date to maturity, requires a lot of space or has a lower price is just not economical for an urban farmer to grow. However, farmers in rural areas aren't nearly as limited with crop selection because they have access to larger tracts of land and don't have as much of a need to turn areas over in a hurry to get something else planted. They also have heavy machinery to manage some crops, like potatoes for example, that are more economically grown with a tractor. But one of the biggest challenges rural farmers face is access to markets. Since they are often removed from the population centers, they don't have the ability to find customers as easily as an urban farmer might. There is a strong case for cooperation here, and could be another great role for urban farmers to play in the community. For more details, see "Small Farm Broker" in Chapter 9.

# Quick Breakdown of Economics

## $19,200 on a 2,400-Square-Foot Yard in 7 Months

Before I delve into the main production systems of a farm and show you how a farm on ¼ of an acre can generate $50,000 and more of seasonal revenue, let me break down the economics of a 2,400-square-foot piece of land. This would be the average size of a front or backyard in a suburban neighborhood in North America.

When we're talking about getting the highest amount of production from a site, we need to specialize the crops that grow there. This means growing crops that grow fast and have a higher value per square foot. On my farm, I group my crops into two categories, Quick and Steady. Quick Crops are those that grow fast (under 60 days), and Steady Crops are those that take over 60 days and are often harvested continuously for a period of time (like tomatoes or kale). I also grow everything in a standard-sized bed. Typically this bed is 30 inches by 25 feet long. In some cases the length can vary, and in some rare cases, the width will vary. But for the sake of planning my farm, I base everything on this standard size. In the 2,400-square-foot area I'm going to describe, I can fit 24 beds.

For this area (what I refer to as a Hi-Rotation (HR) area), I am going to grow Quick Crops specifically. Each bed on the site will be rotated many times throughout the season, sometimes as many as four times. This means that four different crops will be planted in each bed over one growing season.

On a plot like this, I would plant crops such as arugula, carrots, cilantro, salad turnips, lettuce, mizuna, mustard greens, parsley, radishes, scallions and spinach. Notice that carrots are not considered a Quick Crop; they are what I call a Steady Crop, but some crops like this can be put into the mix with a HR area.

I would divide the plot into two segments so that I can access each one with my rototiller throughout the season. This way, if I need to turn over only one or two beds at a time, I can access the plot with ease. One bed in a Hi-Rotation (HR) area can generate $800 because each crop will yield $200 for each rotation.

Following this formula, deciding the outcome of a plot in HR is a simple mathematical calculation: $800 × 24 beds = $19,200. One bed for example might have started as spinach, planted April 1, and it would be ready to harvest by May 15. With spinach, I harvest on average a total of 35 pounds and I sell it at $7 a pound. This 35 pounds is the total harvest, so that might have been two cuts over the course of two weeks. Either way, this harvest is worth $245. Next, I could plant this bed again; I would remove the crop, amend the soil in between plantings or turn it under and plant something else. This time, I would plant a bed of radishes on May 25. This bed would be ready to harvest 31 days later, June 25. Radishes yield on average 75 bunches per bed, and I sell them for $2.50 a bunch; that's $187.50 right there. These would be harvested all at once and sold that week, I call this *cropping out.* The soil on the bed is amended again, and something else gets planted by the beginning of the following week. If this bed were replanted to lettuce on June 29, its first harvest would be July 30. It would be harvested three times over the course of three weeks. With greens I refer to this harvest technique as *Cut and Come Again.* I'd harvest 15 pounds each cut 3 times, a total of 45 pounds at $8 a pound for a total of $360. On August 20 the bed is finished and replanted again August 24. This time I plant arugula, and it's ready in 24 days in the summer. This takes us to September 17, and here we'll get two cuts one week apart for a total of 20 pounds at $10 a pound for another $200. So, in one bed by the end of September, we've grossed $1,005. Even though this number could be higher with some crops, I prefer to use conservative estimates for all my planning, because there are always variables that you can't plan for: weather anomalies, irrigation problems or pest interference, for example. Using this example, it's easy to see how this pattern could be replicated on a broader scale. In most cases, I'm planting beds in groups, like multiple beds of spinach, radishes, lettuce and arugula at once.

PART 2

# A VIABLE FARMING BUSINESS ON ½ ACRE OR LESS

In *The Urban Farmer*, you're going to learn how you can commercially farm on ½ acre of land or less and make a considerable income doing so. The idea is to focus on direct consumer market streams, growing higher value crops and using labor-saving and time-saving techniques I have developed over years putting these techniques into action.

# The Zones of Your Farm and Your Life

When we think about the typical small farm, we usually imagine one piece of land where there would be a home, a couple of greenhouses, an outbuilding for a workshop and storage and maybe a farm gate market stand. In some ways, an urban farm isn't much different, but the ways that all of this infrastructure is laid out is very different. There usually isn't one place where all of these things could be because the land we are using is so much smaller, and often in multiple places. We need to organize the network of land we're using to maximize its efficiency so that we're not wasting too much time in transit, getting from one place to another. First you need to define all of your needs. There are three main considerations:

- Where do you want to be based—city, town or rural area?
- Where will your base of operations be?
- Where will your farm plots be?

## Your City

Your first consideration, and perhaps the most important decision you'll make, is where are you going to be based. It is very tempting to go to places like San Francisco, Vancouver, Portland or even Detroit. I know a lot of people are tempted to go to a place where there is already some urban agriculture happening because it's nice to be in a place where you could have a sense of camaraderie with other farmers, and perhaps use some of the resources already established, whether it's land-share programs, community farms or even access to public money.

What I'm going to say may shock you, but I'd encourage you not to go to anyplace where there is an established urban farming scene. On the other hand, if you can identify a market demand there and you can fill a niche that nobody else is, then by all means go where it's right for you. But be careful about walking into a market that is already saturated. I have seen many farmers do this, and they have a very difficult time actually making money because they're competing with so many other established growers.

Let's be honest here, urban farming isn't really anything new. People have been doing it in some way, shape or form for a hundred years—even in the way I'm talking about, by using front and backyards to farm commercially. There are people in some of those cities I mentioned that have been farming for well over ten years. If you move into places that have urban farms already established, you're story isn't going to be anything unique, and that doesn't really help you.

There's a concept in business called *first mover advantage.* This means that if you are the first person to take a new idea to a place, you'll always be the first one to grab the market share. This will help you considerably in the early years, but it also means that, over time, you'll always be seen as the first one to bring urban farming to your area. That's a huge market advantage. Ultimately, being the first mover makes you a news story, and, when you're a news story, that means free advertising and exposure when you're starting out. Most businesses have to pay for that.

When you're choosing a town or city to farm in, there are a number of factors to consider.

### City Size and Population Density

How spread out is the area? I've been to some towns in the US that are so widely spread out, that one must drive 20 minutes to get anywhere. If you're seriously looking at a place like this, then farming on a single site would be a better option. If the area is a little more concentrated in geography (like a few miles), then being multi-locational can work. I find that cities with a population between 50,000 and 200,000 are the perfect size.

### Demographics and Demand for Local Produce

The size of the city matters only if there is demand for what you're producing. If you have to be the sole person who educates all of your potential customers on the importance of eating local, then you're going to have a far greater uphill battle to get established. Look for places where there may already be a field to fork culture: festivals and events that celebrate local food and farmers, grocery stores like Whole Foods, people who ride bicycles and generally are interested in health conscious living. From what I have seen, these are an urban farmer's target customers.

### Access to Potential Markets

Farmers markets and restaurants are my two major market streams. We will cover farmers markets and restaurants in detail in Part 3, but when you're looking for a place to operate, these outlets for your produce are very important. With farmers markets, look online to see how many markets there are and how many people shop there on a day-to-day basis. Is there a waiting list to get into the market, and how long is their season?

For restaurants, do a similar online search to find how many there might be that use or advertise local food. Do they already use local product, and is there need for more farmers to supply them? Does tourism in the area affect the restaurant industry, and what is the seasonality of that tourism? Being close to your market streams is very important and has to be the greatest advantage for urban farmers. If you have to travel an hour each way to deliver to restaurants or farmers markets each week, you're going to lose your biggest leg up. So, knowing about these markets is a critical element in finding the best city for your farm.

#### *Lot Sizes*

Generally speaking, the more dense the city, the smaller the lots. In my experience, in bigger cities urban farmers will often have more plots of smaller dimensions; in smaller cities, it's the opposite. For backyard farming, I find medium suburban-sized towns have the best of both worlds: good lot sizes and population density. In a multilocational approach, when you have lots of 1,000 square feet or less, they need to be close together; otherwise you can waste a lot of time in transit between them. With larger lots, the space between can be farther. More on this in Part 4.

### Your Home Base

After you've decided where you're going to operate, your next consideration is where your base of operations will be. Your most basic needs on an ongoing basis are going to be determined by the things you do the most often. These are all the things related to packaging, storing and cleaning vegetables—essentially all the tasks that would be done at your base of operations. Ideally this is a place that would be in the center of your network and close to where you live or plan to live in the near future. Wherever you decide to put your base of operations, all other places will stem from this location. It is essentially the heart of your farm and must be central and accessible.

### Your Farm and the Hierarchy of Land

On my farm, I've created two categories in which I organize all my land. I call these Hi-Rotation (HR) and Bi-Rotation (BR) plots. These refer to the amount of times beds are turned over and planted again on the each plot. *Hi-Rotation* areas are plots with constant activity, and each bed is planted many times throughout the season. *Bi-Rotation* areas may be planted only twice in the season. In no instances on my farm do I have beds that have only one crop per season. This is not economical for small land bases, and I don't recommend growing crops with long dates to maturity, such as onions, potatoes, winter squash, melons, garlic or corn.

In a Hi-Rotation area, I primarily plant quick-growing crops, so that each bed would be turned over up to four times—in some cases even more. Sometimes, slower-growing crops (such as carrots and beets) could be grown in these beds, but for the most part, HR plots just grow vegetables with an average days to maturity of 60 days or less. Because farm plots in HR are harvested and replanted on such an ongoing basis, they need to be closer to your home base to reduce travel time. When a farm is multi-locational, it is of the upmost importance that travel time between plots is minimized. During the high season, between July and September, we are visiting the HR plots multiple times per week because almost every day something is either being harvested, turned over or replanted. A bed in a HR area can generate $800 or more of gross revenue per season. That is on average $200 per crop per bed.

Bi-Rotation means that beds are only planted twice during the season. My longest season crops are planted here. Typically they consist of a Primary Crop that is a Steady Crop, and they are either followed by or preceded by another crop. Since BR areas don't need as much constant work, they can be farther away in the plot chain. This is not the case with tomatoes however. As they require constant pruning and harvesting during high season, they should be closer within your network. Some crops that would be the Primary Crop in a BR area would be tomatoes, pattypan squash, carrots, beets and kale. With tomatoes and most summer crops, for example, they don't get set into the ground until mid-May in my climate zone, so I would have time to plant a Quick Crop (something like a cool weather green or radish) before they go in. So the Primary Crop is the later crop as it is preceded by the Quick Crop. Kale for example would be the opposite. Kale would be the Primary, and the Secondary Crop would follow after the kale is pulled out of the ground. I set kale transplants into the ground around the first week of April, and I grow it until midsummer. When the kale comes out, I usually follow it with a crop like fall carrots or beets. In some cases I would do a Quick Crop, but usually something that could be cropped out, like radishes or turnips, so that I don't have to make too many trips to a plot like this, if it's further in the plot chain. A bed in a BR area can generate at least $400 of gross revenue per season. That's $200 per crop per bed.

# Crops Better Suited for the City

It's important to understand that urban farmers cannot and should not try to grow everything. Because urban farmers operate on much smaller land bases than most other farmers, growing some crops is simply not economical for us. There are some other factors such as distance to market, market value, yield per square foot and crop perishability that also come into play. A 19th-century German economist named Johann Heinrich von Thunen proposed the idea of dividing cities into rings of agricultural production in his book *The Isolated State* in 1826.[1] The basis of his thesis was that certain crops should be grown in certain points of a geographical area based on transport cost, land value and distance to market. To highlight this idea in a very obvious way would be to say that it doesn't make economic sense to grow grain in the center of a city. Grain has a long shelf life so it doesn't lose nutritional value through transportation, and it also has a low yield ratio based on the value of land per square foot, making it nearly impossible to make a profit on small land bases. On the other hand, to grow and ship highly perishable

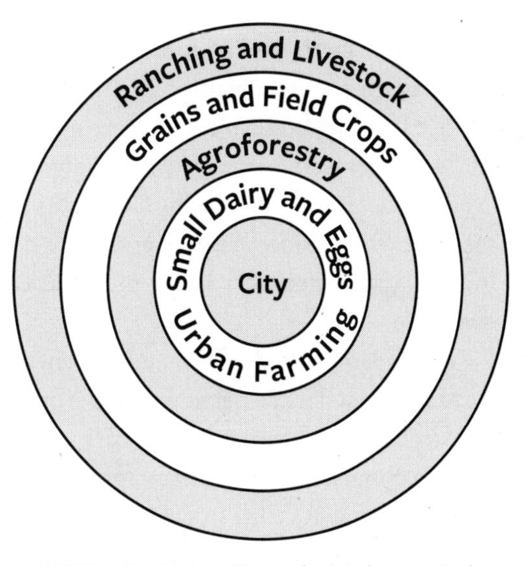


**FIGURE 1.** The Isolated State depicts how agriculture can be built into human settlements. Credit: Curtis Stone.

salad greens from a 1,000 miles away is not economical either. Consider all the factors that make fresh greens valuable: they have a short shelf life, a high yield per square foot ratio and a short date to maturity. This makes the idea of growing this crop close to where it's consumed a good choice. Short date to maturity is very important for the urban farmer: these crops can be replanted many times over to assure the maximum amount of production based on the yield-per-square-foot ratio.

The Cuban Special Period is another great example of this fundamental practice. As the Soviet Union was collapsing in the 1990s, Cuba lost access to most of its fossil fuels and conventional fertilizers used for agriculture due to an embargo from the US government. During this time, Cubans were starving, and drastic measures had to be put in place to feed people fast. Crops like salad greens were planted in the inner city because they could feed people quickly and were harvestable close to where people lived. Such innovations are why Cuba is still seen as a world leader in urban agriculture. In many cases, necessity is the mother of invention.

Von Thunen's and Cuba's approaches are fundamental when deciding what crops are most suited to urban agriculture production. Similar principles also apply when we are looking at crops to grow on an urban farm, but they are a little different when taking into account what makes the most sense to grow from logistical and financial sides. In my own experience, there are five characteristics that make a crop well suited for growing on a small land base in a city. I summarize these characteristics as a *Crop Value Rating* (CVR). CVR works by assigning each crop you want to grow one point (out of five) for each characteristic it has:

1. Shorter days to maturity (60 days or less)
2. High yield per linear foot (½ pound per linear foot, per 30 inch by 25-foot bed)
3. Higher price per pound (minimum $4 per pound)
4. Long harvest period (4 month minimum)
5. Popularity (high demand, low market saturation

Let's look at each category and what it means.

### *1. Shorter Days to Maturity*

Shorter days to maturity (DTM) mean that the crop grows and is ready to harvest within 60 days or less. Radishes are 28 days, so they'd score here. Spinach is 45, so it scores. Tomatoes are often 70 days or more, so they don't score here. However, I don't grow anything on my farm that has more than 70 days to maturity from the time it is set into the ground. This means that I will grow only a tomato variety that is ready in about ten weeks from transplanting; the same goes for carrots, beets or anything else that would have a longer DTM. I always try to grow the fastest varieties, no matter what the crop is.

#### *2. High Yield per Linear Foot*

Consider how much yield will come from one linear foot of bed. For example, a cabbage takes around 80 days to mature. When the plant is at 75% maturity, it takes up around two square feet in a bed; all of that space will yield only one item that is sold once, at a relatively low price. Compare cabbage to yield from radishes. Radishes are ready in 28 days, and in the same space as one cabbage, I could harvest eight bunches. The general idea is that you get maximum value from space.

#### *3. Higher Price per Pound*

Generally speaking, I don't grow anything that sells for less than $4 per pound. My lowest value crop to meet this requirement would be cherry tomatoes at $4 per pound. Some of my tomatoes score lower, but that doesn't mean I won't grow them; it just means if I had to cut a crop from my list, they might be the first to go. Kale or radishes at $5 per pound also meet other criteria: long seasonality and high yield per square foot. The highest price per pound would be microgreens at $20 per pound. So, in order to meet this price criteria alone, a product must be sold for at least $4 per pound. This does not mean don't grow anything that sells for less than $4; it just means that, if you do, make sure that crop meets some of the other criteria as well.

#### *4. Long Harvest Period*

This can mean either of two things:

- It's a crop I can keep planting and replanting for a minimum four-month period.
- I can keep harvesting the same crop for four months.

Radishes, for example, I can keep planting all season, so they have a long season in that regard. A crop like kale I can harvest for most months of the year, except during the heat of summer. Tomatoes don't have a long harvest period, but I'm going to grow only varieties that have the longest period I can find, such as indeterminate types that will bear fruit for many months.

#### *5. Popularity*

High demand and low market saturation are perhaps the single most important of characteristics of them all. You can grow high-value and quick-growing crops until you're blue in the face, but if nobody knows what they are or wants to buy them, they're not worth anything. I learned this the hard way: one year I drastically scaled up my microgreens production. I quadrupled my output from one year to the next, and saturated the market. I was no longer able to sell them all. They also became an item that other farmers started to grow because of their high value. Once the market becomes saturated, pretty quickly they became passé to the chefs. That left me sitting on a lot of product that I couldn't move.

Kale is a perfect example of a crop that might not be the highest value per pound—but it's very popular and I never have to work too hard to sell it. It might not have all of the criteria for our CVR, but it scores overall because of its popularity.

### *Some Examples to Illustrate CVR*

Let's look at three crops that all have a CVR of from three to five. (Keep in mind that this rating maybe different for the same crops in differing areas, based on what demand there is for it; is the market saturated or is there high demand?) In my area, a crop like spinach rates 5/5 on the CVR. Spinach has short days to maturity (45 or less), a high yield per linear foot (1.4 pounds per foot), a higher price per pound ($7 per pound); it has a long overall harvest period (10 months), and it's a common vegetable that is popular because of its many uses. Cherry tomatoes on the other hand score three out of five. They have a high yield per linear foot at 7 pounds; they can command a good price at $4 per pound; and they are very popular with both of my market streams. Beets score four on the scale. They meet all the requirements except a short DTM: they take 60–70 days to mature.

Using CVR rating, you can establish a system in which crop choices are rated based on logistics and economics. Most crops on my farm score a four out of five on average, with a few threes and fives. The principle idea is that, the smaller the land base, the higher you want the majority of your crops to score on the CVR scale, to make up for the loss of land. The larger the land base, the more crops you can include that score lower. If you were to grow solely on ¼ acre of land, you should grow only crops that score fours and fives. On any smaller farm, grow only fives. On ½ acre or more, include more variety in order to capture more market share: grow threes, fours and fives, and perhaps even some twos. On a farm this size, you may start to grow longer-season crops that are not covered in this book.

### Cash is King—Quick Turnover

Following these basic principles, we can easily see why some crops are just better suited for urban production. The other factor that is critical here is *cash flow*. Fast-growing annual vegetables have a fast payoff. They are planted and, in some cases, ready for harvest in 30 days. Planting perennial crops often takes years to see a return. Quick turnover is particularly important when the farmer doesn't own the land. Waiting for a three-year return on investment for an asparagus crop, when you might have access to that land for only three years, is not a sound business decision. In all these cases, we must come back to basic economics and sound business practices.

Many people want to grow vegetables that they like or perhaps because they have some ideology, such as health or environmental benefit, based around them. It's good to have ideology and an ethical stance on things, but if your ideology makes you go broke, then nobody wins. Don't abandon your ideology; just keep it in your back pocket and be practical. As you make small successes, bring some ideology back and implement it incrementally. If you start with too many grandiose ideas, it will be hard to get started. Don't think that you're selling out because you can't use all the ultimate ideas of sustainability at once. It's good to have huge ideas, but make small steps to get there. This system is about going from A to B. What the ultimate form of sustainable agriculture looks like, I'm not totally sure, but we're taking steps to get there, and we figure it out along the way. The important thing is: first, you need to make a living at this. Otherwise, you're working somewhere else and are doing this on the weekend. You must have cash flow, and this is why I grow the crops that I do.

## Quick and Steady Crops

All the crops on my urban farm are divided into two categories, and again, these crops are all based on what makes the most logical and economic sense to grow in a city. *Quick Crops* grow fast and are often, but not always, harvested at once to free up space to be planted again. *Steady Crops* are a little slower but are harvested over a long period (most often harvested on a weekly basis throughout the season). In no cases do I grow any long-season crops such as onions, potatoes, cabbage, winter squash, melons, corn or garlic. Crops such as broccoli, cauliflower, brussels sprouts, beans and peas are not great for my farm either. And even though some of these crops may have shorter dates to maturity, they have a very low yield per linear square foot. I don't recommend that any urban farmers grow crops like this for commercial production; however, to grow them for your own use is another story. Crop specifics are covered in detail in Parts 9 and 10.

### *Quick Crops*

Quick Crops are ready for harvest in 60 days or less. For the most part, I replant all crops in this category on an almost weekly basis. When I plan my farm, I don't plan where all my Quick Crops will go. I simply decide what areas will be in Hi-Rotation and that's it. I leave the exact details of these crops to bend to the demands of the market, and the exact amount that gets planted on a weekly basis will change from season to season. This is partly how I maximize the economic output of my farm. If something isn't selling, it gets terminated and something else gets planted in its place immediately.

### *Steady Crops*

Steady Crops are those with a date to maturity of 60 days or more, and they are called Steady because they are often steadily harvested. Tomatoes, kale and summer squash, for example, are picked weekly during the high season and constantly regenerate, whereas sometimes carrots can be totally cropped out at once or harvested a few rows at a time over a week or so. In some cases, beets can be thinned out over the course of a few weeks.

## All Season Production

Part of what makes my farm profitable is having a steady level of production for the entire season. My season here in Kelowna, BC, is 30 weeks long on average, and the key here is to maintain a strong income for the entire period. This is not like the typical farming season, where the shoulder seasons are low and the summer is high. This is why you'll notice that a lot of the crops I grow are spring crops. The majority of vegetables I grow and the base of my farm income come from crops that I can grow in almost all seasons. This is very important because it allows for a strong foundation of cash flow all year. Granted, there is a small amount of a seasonal spike in the summer which is inevitable for any farm. But the main idea is to have strong weekly sales all season, and I choose the kinds of crops I do based on this idea.

### *Winter Crops*

With some really simple season-extension techniques, it is possible to have some field production all winter, even in some Canadian winters. There are three types of greens I can have almost all winter in unheated greenhouses or low poly tunnels: lettuce, spinach and kale. It's also possible to have carrots in the ground all winter. I use a technique called overwintering to achieve all of this, and I go into more detail in Part 8. Growing microgreens indoors is also a great way to have production over the winter. You need to spend a little on infrastructure, but it's possible to make $2,000 per week in a 400-square-foot area growing microgreens: see Part 8.

### *Summer Crops*

It is important to have a strong summer production for two main reasons:

1. You need to satisfy demand as it occurs.
2. Spring crop production will slow down in summer heat.

Yields for some crops will drop in very hot climates, and this happens in my area. We are in a high desert here and experience daytime temperatures of 104°F some days. Crops like arugula, lettuce and mustard will yield up to 50% less than in the spring, but they will grow faster, so it can be a bit of a trade-off. Also, I cannot grow crops like spinach and bok choy at all during summer. They will bolt too fast and sometimes not even germinate at all. For the summer, I make sure that I have strong production of indeterminate tomatoes, summer squash, some peppers and a lot of carrots ready to go as the spring crops start to taper off. The nice thing about the summer crops that I grow is that in their early stages it's possible to interplant them with some spring crops. This way you can maximize the use of the land while those summer crops are not yet producing. See Part 9.

### *Specialty Crops*

There are some cases where I will grow a certain crop for a specific chef customer. This is something that is planned long before, and is usually for some type of one-off event. It's not something I would suggest doing a lot of, and on my farm, I do this only for chefs that I've worked with for a year or two so that we have a good rapport. These types of crops are most commonly a special microgreen, some kind of baby vegetable like a super small dime-sized radish or turnip, a tiny bok choy or some kind of baby green or herb. More on this in Part 3.

# Introduction to Urban Infrastructure

For the most part, infrastructure on my farm is simple, inexpensive, modular and often very DIY. Besides major purchases like walk-in coolers, transportation, rototiller and a nursery, most things can be purchased used or built yourself. I started my farm on $7,000, and that covered all my major purchases, tools and basic infrastructure. In Part 6, I cover these items in greater detail.

# Start-Up Farm Models

Below are five different business models that can work for a commercial urban farm operation farming on ½ acre or less. I have run my farm at various sizes and scales over the years, and a lot of the models in this part are based on what I have tried. Some (like the Small Farm Broker) are ideas that I think would work but have not yet tried. I have done some brokering before and still do a little bit, but this is a model which I think would be a great option for someone wanting to operate on a very small land base, like ¼ acre or less, and still have the potential of pulling in a six-figure gross income.

## How to Start on Less than $10,000

One quarter acre of land or less is the right amount to start with if you don't have any previous experience in farming. Taking on too much land is the number one mistake I see new farmers make. On ¼ acre, you have the potential of making $50,000 from the land itself, but if you incorporate some greenhouses or indoor microgreens, you could make that number considerably higher; this all mostly depends on your market streams. Understanding your market is the key to success in farming. I will revert to this many times in *The Urban Farmer.*

In order to spend less money on start-up, you'll need to spend more time looking for deals on your major investments to save cash. If you can give yourself six months prior to starting, like I did, that should be enough time to build the infrastructure you need, prep some land, and look for the best deals on good used equipment. Depending on how populated the area you're in is, you may have to do some traveling to find the best bargains in used equipment. Using sites like craigslist and Kijiji, I found a lot of great deals, but I sometimes had to drive four hours to pick them up. It was all worth it though. I purchased a BCS tiller with three implements for $1,000, and I bought my first walk-in cooler for $1,000. If I had bought both of those items new, I would have spent $8,000 more. The point here is to shop around. Also, use craigslist or other sites to post what you're looking for. I found my BCS because I made a post saying that I was looking for one. The $7,000 I spent in my first season covered all of my major investments, seed, tools, irrigation and fertilizer.

If you want to achieve ambitious revenue from ¼ acre or less, then you need to have access to high-end restaurants and good farmers markets. You will also need to specialize in the crops that give you the highest return on the smallest amount of land, in the least amount of turnover time. This means a little less diversity in crop selection.

On the other hand, if you're going to farm ½ acre or more, you'll need to broaden your market reach, unless of course you have access to a lot of restaurant market streams. Unless I could act as a broker, I would consider operating a CSA only on ½ acre or more. The advantage of catering to restaurant markets is that you can grow large quantities of vegetables that have high margins. For example, I grow a lot of baby root vegetables like radishes, because some of my customers will go through up to a hundred bunches a week. I pretty much exclusively grow them for restaurants. I can sell up to 200 pounds a week to all of my clients on a weekly basis, but there's no way I could sell that many at the market, or even in a CSA. I'm lucky to sell 20 bunches on a good market day. You learn over time what sells and where. Some items (like swiss chard and kale) I sell better at the market than to my restaurants. It takes a little time to learn what products do the best in each area; this is why taking notes on spreadsheets is so important. You need to continuously update this information and leverage it over time to make your production and sales as effective as they can be.

The models below are scalable. For example, it would be possible to start as a part-time farmer, growing in your own backyard, and then, once you get some experience and feel comfortable quitting your day job to pursue farming full time, scale up to the Microfarm or Small Farm after a year. From there, you can continue to scale it as you see fit.

The biggest challenge when scaling up is finding the right help, getting people who are willing to work at least close to your level. Anyone who is an employer will tell you that you'll never find anyone who will work as hard as you do, and for the most part, they're right. The key is finding what level of output is acceptable. I have found that the best employee will work at 85% of my output level—and that's really good, like best-case scenario. You have to accept the fact that nobody will work as hard as you because they don't have the incentive that the farm owner does. If you can create incentives such as pay bonuses based on production or revenue targets, then I think you can increase that percentage to some degree.

### $21,600 Part-Time Farmer in Your Own Backyard

An ideal situation for anyone would be to keep your day job for a year or two, perhaps scale your hours back a little bit and run a farm on a part-time basis. This way you can give yourself some comfort and security in knowing that you don't have to go in head first. I will say, though, that jumping in head first makes you learn faster because you have a lot more skin in the game. But I understand that not everyone is willing to take that big a risk at first.

This model is for people who see themselves in that position. One tenth of an acre—around 36 beds at 30 inches by 25 feet on one piece of property—would be an ideal size. Ideally, your own front and backyards or maybe somewhere really close to home (like right in your own neighborhood) would be your farm. If you were to run this farm with all HR beds growing Quick Crops, it could gross $28,800 in a 30-week season. But, I would recommend diversifying your crops to give yourself experience with growing a variety of different things. Thirty-six beds at half BR and half HR could still generate $21,600 in a 30-week season. The outcome for this model is less about maximizing profit and more about learning systems so that it can be scaled up at a later date.

To run a farm in this way, you have to focus on market streams that are less risky and take less time in customer service: that's primarily farmers markets. Because most markets run on weekends, this makes it possible for someone with a Monday-to-Friday job to do a lot of work on the weekends and evenings and successfully operate this model. It might be possible to try and market to restaurants a little bit, but I suggest just small owner-operator types who you feel confident that you could supply. In this case you're looking for places that will use small amounts, like an order of $60 per week, for example. Also make sure they can be flexible with what they get, because when you're learning, you will find sometimes things may be a little inconsistent. The challenge with catering to chefs is that they require more on the customer service end. Text messages and phone calls midweek are very common, and if you're not available for those, it could make it more difficult to engage with those customers.

Realistically, you'd have to scale back your hours at your main job, for example, Monday through Wednesday 9–5, and then working an hour or two in the evenings or early mornings before work. And then Thursday 12–5, so you have time in the morning to harvest what you need for market. You'd need to have Fridays almost entirely off, or a really short day of a couple hours, because if you were to sell at a Saturday market, you'd need most of Friday to process and package all of your product.

I have seen small families with children and couples make this type of farm work quite well. The more collective support you have, the more options you have for when the farm work can get done. You'll need to be able to commit at least 20 hours a week plus a half day at a farmers market in order to make this model work.

### $58,800 Microfarm on 1/10 Acre

Farming on 1/10 acre could be running a farming business at your own home in the suburbs or urban lot. If you had a front and backyard that was a total of 4,356 square feet, that could be enough land to actually run a farming business from home. With this model, in order to maximize profits, you must only focus on Quick Crops and Hi-Rotation; this means growing only crops like greens, radishes, turnips and some herbs, and turning those beds over at least four times in the season. At this size of farm, you would have around 36 total beds and about 10 inches for the walkways between the beds. If all these beds were in HR, the income potential is $28,800 from just the field crops alone in a 30-week season. This number could get higher by using greenhouses for the field crops and microgreens. If you could produce and sell 50 flats of micros per week at $20 per flat, it would be possible to gross $58,800 from a lot this size. Also, if your season is more than 30 weeks (which is the average for North America), the possibilities are greater.

The risk with this model is that your products and market streams are very specialized. So, your success is based on being able to cater to some very niche markets and being able to move all the ten or so crop varieties you grow. A farm like this might work well in a larger city area where you have access to some very high-end restaurant markets that are willing to buy only specific things from you. In no way would you be a one stop shop for anyone, and that means that you have to accept the fact that you're catering only to the specific needs of your clientele—and not even their basic ones. The other risk here is that because the scale is small, your restaurant customer base will also be pretty small, meaning that you are reliant on a small handful of customers to support you. If you were to lose some of those customers, then your income could be compromised very quickly. These are some of the risks involved in primarily focusing on the restaurant markets. Sometimes if a certain chef leaves, then you have to go back and try to establish a relationship with the new one. I have had customers in the past that spent $1,000 per week, and then the next season the restaurants changed concepts and chefs, and they were basically gone as customers. Some diversification with customers is important, but it's even more important to build good relationships with your clients and constantly keep your eyes and ears open when communicating with them.

Running a farm at this scale could be a full-time job for one person working at least 40 hours a week. The microgreen production would be close to 15 hours per week alone, and the rest of the time would be spent on the field crops, processing and delivering. If you were able to only sell to restaurants and not bother selling at a farmers market, then that would save you an entire day's work standing at a market booth.

The crops I recommend growing for this model are arugula, cilantro, baby dill, salad turnips, baby lettuce, mustard greens, radishes, baby red russian kale and baby spinach. All the greens I mentioned, except spinach, should be mixed together for a variety of salad mixes, and the spinach and arugula could also be sold on their own. Try different mix combinations to market as different products. Mustard and arugula together form a "spicy mix," for example. Grow microgreens such pea shoots, sun shoots, radish shoots and some specialty microgreens such as purple opal basil, cilantro and anything with bright color. Pea shoots and sun shoots are common within the health food community, and this makes it possible to produce a lot of those and market them to that demographic. Radish shoots and specialty microgreens are popular in restaurant markets.

### $87,000 Small Farm on ¼ Acre

Farming ¼ acre is the perfect place to start if you are new to farming, and you are prepared to make this your full-time job. To achieve the highest potential gross income from this model, you will most likely have to employ at least one other person, part-time or full-time, depending on how much time you are willing to commit.

In this model focus on Quick Crops and Hi-Rotation areas. No Bi-Rotation crops like tomatoes or summer squash would be grown. However, it would be possible to have up to four beds of kale in here without affecting your income negatively. On ¼ acre, you would have around 90 beds with around ten-inch walkways between the beds. If you're growing in a cold climate, I'd suggest using either poly low tunnels or high tunnels to push the season longer. With very few season extension techniques, you should be able to market for 30 weeks if you are in most places in North America except for climate zones of 5 or lower. In those climates, a 20-week season would be reasonable. Ninety beds in HR can generate $72,000 in 30 weeks or $2,400 per week on average. If you were to add some microgreen production to this model, you could bring in a considerable amount more income. If you just targeted 25 flats per week at $20 per flat for 30 weeks, you could bring the total gross up to $87,000, and if you were to be a little more ambitious and produce 50 flats per week for 30 weeks, it would be possible to gross $102,000 from this ¼ acre.

The crops I'd recommend growing are turnips, bok choy and radish bunches, arugula, baby red russian kale, spinach for loose greens and lettuce, mizuna, mustard and tatsoi for greens mixes. You could also add red russian and arugula to those mixes as well. Bunch herbs such as baby dill, cilantro and parsley for sale. You could also grow kale for bunches to add a little more variety. Grow microgreens like pea shoots, sun shoots, radish shoots and some specialty varieties for restaurants and farmers markets. The market streams I would encourage are a balance of farmers markets and restaurants, and look for the possibility of providing some niche products into another farmer's CSA. I wouldn't recommend running your own CSA at this size, because there is not enough variety in your products.

The advantage of this model is that you are growing a small diversity of crops, which offers you more marketability in both farmers markets and restaurants. So, this scale of operation would allow you to broaden your customer base a little more easily.

### $123,000 Semi-Diversified on ½ Acre

One half acre is the largest amount of land I'm going to propose farming. Anything beyond this, you're getting more into rural farm plans that people like Jean-Martin Fortier or Eliot Coleman put forth in their books.[1] Half an acre is a lot of land for an urban farm; in most cases, if you're focusing on high value crops, you'll most likely exceed what your markets will bear, so you'll have to offer a wider variety of products. Also, with a farm this size, you're going to have to have an employee or two. This model is not recommended for anyone new to farming.

Operating ½ acre would be for those who already have a season or two under their belt. It would be possible for a farm this size to be operated by two full-time people if they were co-owners. You might have some temporary help in the summer months for help with market prep on Friday, but for the most part, two experienced people could farm ½ acre and still balance life and workflow.

This much land would allow for 180 beds. I'd recommend farming an equal amount of HR and BR beds and all of the crops I list in Part 10, with some indoor or greenhouse microgreens as well. With this much production and crop diversity you have the ability to market to a fairly wide customer base, including restaurants, farmers markets, and perhaps to operate a small CSA program. Even at ½ acre, the primary value a CSA program offers to an urban farmer is the up-front cash at the beginning of the season, not the income it brings. With a larger farm, you will have higher overhead costs, especially start-up costs. Your fixed costs for seed, labor, fertilizer and transport will be considerably higher, and you will have to be very careful to make sure that expanding to this size of farm is actually worth it. If you can operate at this size and still pursue high-end markets, then any extra costs will be absorbed by your extra profits. However, if you have to engage a lot more lower-yielding market streams like CSAs and more days at the farmers market, you might still be making the same net profit at the end of the day. The key to expanding a farm is understanding what your market demand is and if there is room to grow. If you are saturating your markets, then expanding your operation is pointless. If, however, you're selling out at the farmers market every week and you can't seem to grow enough for your restaurants, then those are obvious signs that you can expand to meet demand.

I find the best place to be is where you're almost meeting the demand, but not quite. It's better to sell everything every single week than to not sell 15% of your product. I'd rather be 5% short than have 5% left over. Every time you go home with product from the market or are producing too much, then you are working for free, and that really cuts into your lifestyle and bottom line. Part of what has allowed me to have such a great work/life balance is the fact that I pretty much sell everything I grow. This way, I'm never working without getting paid.

# PART 3

# THE BUSINESS OF URBAN FARMING

When I was growing up, I would spend the weekends with my dad who, when I was very young, worked as a traveling salesman. He took my little brother and me on the road for his weekend trips around southern BC. He'd take us into stores with him and introduce us to his customers, and we'd watch him make deals and shoot the breeze with people from all walks of life. I remember always being fascinated by all the different and sometimes eccentric characters we met. Later on into my teens, my dad bought a small fish and chips restaurant, and a year or so later expanded into a couple of nearby towns. My first job was at 14, working for my dad in his restaurants. I would work at the three different locations as a dishwasher, and he'd take me wherever he went. At a very young age, I got a sense of what it was like to be an entrepreneur, and for this I'm ever grateful to my dad for teaching some fundamental things to me.

Today, I count myself lucky to have been brought up around all that. My dad didn't have much; when he was running the restaurant, he had far more debt than was manageable. I learned through watching him that if your expenses outweigh your income, you're in danger.

In my dad's later years, he pursued a dream he had had since he was really young. He became a barber and opened his own tiny shop. He worked for himself all the way until he retired. He didn't make tons of money doing this, but he was happy. He felt a sense of purpose with it. I have always admired him for doing exactly the opposite of what society tells us to do: you must grow and expand constantly! He found happiness by doing the opposite.

I see so many farmers struggle because they fail to see their farms as businesses. Often they carry so much ideology with them going in that they think there is only one way to do something, not recognizing that there are some ideas that are universal to all businesses, small farms included. All my life, my father hammered some basic principles about running a business into my head, and these have helped me beyond measure. I still hold these as critical to running a business or a farm:

1. If your expenses outweigh your income, you need to change something.
2. When you say you're going to do something, you do it. A handshake is a done deal.
3. A deal is not a deal unless both parties are absolutely happy with it.
4. If you make a mistake with a customer, you make it right by making it better than right.

# Starting Small

Many businesses fail in their first couple of years, and one of the main reasons is that they go in with too many overhead expenses. Farming is no exception to this. In fact, it's often more common. Unfortunately a lot of farmers in the west, instead of going out of business, just keep operating under credit. In a world of cheap money, this has allowed us to perpetuate many systems that in any other circumstance would be totally unsustainable. Small overhead and start-up costs are what give the small farmer and urban farmer a huge advantage getting started.

## Keeping Costs Low

When you look at the numbers based on the size, my farm is lucrative because my overhead expenses are very low and my start-up was considerably lower than average. The fact that I didn't have to pay for any land in the beginning was critical to my success. My transport, input and labor costs were negligible, compared to other businesses operating on the same scale. Transport was cheap because I operated the farm entirely by pedal power when I started, and I still do mostly everything by bike today. My input costs haven't really changed much over the years; they are primarily compost that I buy from a local business and an organic-based fertilizer from a BC company. Each bed I plant costs me around $5 in inputs, most of the time less. Labor has been cheap because I mostly do the major farm work myself. I have some neighbors who help with some specific tasks like tying tomatoes and packing on Friday for market in exchange for veggies. Outside labor works out to be around 10 hours a week from a few people in the neighborhood. It's a good deal for them, and they enjoy doing the work. I have employed many people over the years but in the year that was most lucrative, I had one part-time person working 16 hours a week to help with some very specific tasks such as harvesting and washing bunched veggies. I did most of the production work, such as planting, bed rotation and maintenance myself. Because of the size of the operation, this worked very well. As I have stated before, as you get close to ½ acre, you will most likely need two full-time people to operate the farm, with some extra help on market prep days.

# Market Streams

It's always a good idea to be diversified. Any economist will tell you that. Farming is no different. You want to have a selection of different market streams that you can sell through, and there is really one style of selling that is most relevant to small and urban farmers: direct consumer marketing—selling directly to the end user, without anyone in the middle. In some cases it can make sense to pursue "middleman" types of customers, but for the most part, you are a retailer (the primary seller). The three primary market streams for most small farmers are farmers markets, restaurants and CSAs, though we will look at a few other options as well.

In each of these market streams there are different ways that your product is marketed. For example, with farmers markets and CSAs, you're selling product in small unit sizes at a premium price. That price reflects the extra time it takes to assemble and sell. Units like small bags and bunches are what's common here. For restaurants, you're selling product in case lots (by the pound or by the box), and you are selling at a slightly lower price than you would at the market to reflect the reduced amount of labor that goes into those orders—and the fact that the customers are buying in volume. When you're selling through distributors or retail outlets, you're preparing the product as for a farmers market (by the bunch or bag), but you're producing high volumes so you're still offering a price discount based on volume. We'll look at all these market streams in closer detail.

## Farmers Markets

Farmers markets are the easiest and fastest place to start selling your produce. Some markets will have waiting lists and a specific time of the season they are open, so you need to be aware of these as you think about where to sell your produce. When you are starting out, the farmers market should be the first point of sale to consider. Market customers can be very forgiving to newbies. If you show up with 100 bags of spinach one week and only have 75 the next week, nobody is really going to notice. In fact, sometimes that can help you. Humans have a tendency to react to the thought of scarcity, so if you have less spinach this week, your customers will probably just assume you are selling out quickly, and they'll show up earlier and buy more next week.

The other aspect I love about markets is that they're a great way to debut new products; you get a firsthand response to certain things. It's like having your own marketing focus group, giving you immediate feedback on your product. If you pay close attention to the responses, you can learn a lot about the needs and demands of people in your community very quickly. Markets are a great way to meet people as well. I met most of the chefs I still work with today at the farmers market, and most of the CSA customers I had, when I ran one, I also met there.

Hands down, my favorite part of selling at farmers markets is the people. You get to sell the fruits of your labor, which you have put so much time, care and hard work into, to the kinds of people who value what you do. It's like having all the kinds of people you would want to hang out with right there supporting you. How many people can say that at the end of a workday? People at a farmers market want to hear your story so, in a way, you become your own brand. This is important, because it's what makes you unique, and that's part of the reason people come to the market in the first place. It's the place where people go to experience culture and diversity. All you have to be is yourself—but you have to be polite; you have to be outgoing; and you must smile! If you're not comfortable with all that, then fake it till you make it. If you don't feel like smiling that day, then force it. After ten people smile at you, you won't have to fake it any more. You'll be looking for that connection.

It's important to understand the value you bring to the table at a farmers market. You're not just selling a product, you're allowing people an opportunity to be part of a movement and an idea. Urban farming is good news, and people love the idea of greening the city and bringing food and knowledge about food to them. You're not only selling vegetables, you're also selling an experience. Always remember that. When people buy produce from my stand, they get my undivided attention for the time they're there. I'm constantly answering questions about gardening and giving them tips on how to prepare food. You don't get that at a grocery store, and it's good for you to understand how valuable to a community you are. See photo #1 (farmers market) in the photo insert section.

### *What to Sell at the Market and How to Sell It*

I rarely sell product by the pound to be weighed at market. There are some items that lend themselves to that, but for the most part, I pre-bag or bunch everything before it's brought to market. Most of the items I sell at market are sold under a simple blanket pricing system. There are many ways you can price things at your market booth, but the basic idea is to keep things simple. Round everything off to the dollar ($2, $3, $4, $5). Avoid numbers like $2.50, $2.75. Simple change-making will make transaction times a lot faster; that's better for you and better for the customer.

Generally speaking, it's better to have larger units at a higher price than to have smaller units at lower prices. This is just basic economics: if it takes the same amount of labor to pack a $3 bag of greens as it would a $2 bag, then $3 is better. But you need to take into account how much the average person will use on a weekly basis, because by that same logic you'd assume that a $5 bag would be better. In some cases it is, but I find that most people don't want a $5 bag—though giving them the option wouldn't be a bad idea.

At my market booth, each item is $3 or you can purchase two for $5. This goes for all bunches of radishes, kale and chard, as well as bags of greens, pints of cherry tomatoes and bags of microgreens. However, this system doesn't work for everything: for example, small bunches of herbs, large heirloom tomatoes or larger bags of items like carrots or spinach. To make things as simple as possible at my booth, I will place all the items that work within that price tier of $3 each or two for $5 in one area, and everything that is separately priced somewhere

**FIGURE 2:** Market Packing Legend. Prices that show $2.50 are $3, or two for $5.

| Crop | Package | Weight | Price per Unit |
|---|---|---|---|
| Arugula | bag | 4 oz | $2.50 |
| Basil | bag | 4 oz | $2.50 |
| Beets | bunch | 12 oz | $3.00 |
| Bok choy | bunch | 12 oz | $2.50 |
| Braising mix | bag | 4 oz | $2.50 |
| Carrots (baby and rainbow) | bag | 12 oz | $3.00 |
| Cilantro | bunch | 2 oz | $2.00 |
| Kale | bunch | 8 oz | $2.50 |
| Parsley | bunch | 2 oz | $2.00 |
| Pattypans (baby size) | basket | 10 oz | $2.50 |
| Pattypans (medium size) | basket | 1 pound | $2.50 |
| Pea shoots | clam | 2 oz | $2.50 |
| Peppers | basket | 8 oz | $2.50 |
| Radishes | bunch | 8 oz | $2.50 |
| Radish shoots | clam | 2 oz | $2.50 |
| Red Russian | bag | 4 oz | $2.50 |
| Scallions | bunch | 4 oz | $2.00 |
| Spinach | bag | 6 oz | $2.50 |
| Spinach (large) | bag | 14 oz | $5.00 |
| Spring mix | bag | 4 oz | $2.50 |
| Spring mix (large) | bag | 9 oz | $5.00 |
| Sun shoots | clam | 2 oz | $2.50 |
| Swiss chard | bunch | 8 oz | $2.50 |
| Tomatoes (cherry) | basket | 10 oz | $2.50 |
| Tomatoes (large heirloom) | single | by lb | $2.50 |
| Tomatoes (medium size) | basket | 12 oz | $3.00 |
| Turnip (Hakurei) | bunch | 8 oz | $2.50 |
| Zucchini (baby) | basket | 10 oz | $2.50 |
| Zucchini (medium size) | basket | 10 oz | $2.50 |

else. It's all about making it simple to clearly see where things are. I sell all bunched herbs (like baby dill, cilantro and parsley) at $2 each, and I don't offer any discounts there. I also sell small scallion bunches at $2 as well. Greens such as spring mix, arugula and spicy mix are sold in four-ounce bags at $3 and two for $5. For spinach, which I find is heavier than lettuce and arugula, I will sell a six-ounce bag at $3. All these prices are primarily based on my production. During the spring time, a spinach bed will have high yields, so I extend that value to the customer. At times that I'm producing a lot of greens, I will offer a $5 bag of greens that weighs nine ounces, offering a better deal to encourage volume purchases.

### *Tips for Selling at the Market*

**1. Work on Your Customer Service Skills**

Learn to be polite and friendly with people. The beauty about the market is that people generally want to chat; it's part of the reason they go there. You're going to need to get used to chatting a lot about your farm and telling your story over and over again. The more you do it, the better you get at it. More importantly, identify who your most supportive customers are. Who are the people who come to your stand every week? It's often said that 80% of your business comes from 20% of your customers. (This is true for a lot of things, and this is known as the Pareto Principle.[1]) Identify that 20% as early as you can, and learn everything you can about them. Treat them like friends. Remember their names, what they do and what they're passionate about. Welcome them every time they come to your stand, and make conversation. Only do this if you can be absolutely sincere. In no way should you be fake about this. You have to enjoy it; otherwise it'll have very little meaning. But, what will happen over time is that those people who are consistently buying from you will become your biggest advocates. Once you establish a strong rapport with these people, they'll sing your praise and keep bringing their friends to your booth; they'll introduce their friends to you as if you were family. That's what's it's all about.

**2. Anytime a Customer Complains about Something, Don't Hesitate to Compensate Them**

Give them whatever they purchased before plus something extra. Don't even hesitate if you think they maybe exaggerating or just being a complainer. Hopefully it's not something that is consistent, but that's just life to a certain degree. Every time someone comes to you with a problem, you are presented with an opportunity to make things better. Never hesitate to take that opportunity! This is something that doesn't happen very often for me, maybe a couple of times a year. When you show a person you care about them and what they think, and you're willing to make it right, you're extending gratitude towards them. When someone comes with a problem and they're probably expecting attitude in return, I always make it right—and they always leave with a smile on their face. The next time they come back, they bring a friend. This is why I never hesitate with these situations, because either that person will never come back, or they'll come back with someone else. It's up to you, but the logic behind being generous and humble is pretty simple.

### 3. Create the Illusion of Abundance and Be Creative with your Limited Space

There's an old saying: "Pile it high, watch it fly." People respond to the look of abundance. It's kind of like how you'll see a mountain of soda pop at the grocery store. People are drawn to the look of a pile. The trick for urban farmers is that we don't really have mountains of product; we have a variety, but usually in small amounts. So, you need to manipulate your space to create the illusion of abundance. I do this by pushing different products close together and stacking them as much as I can. I try to use as much vertical space as I can as well.

### 4. Always Have Promotional Material with You at Your Stand

When you are starting out, always keep flyers, business cards and any other type of promotional material close by. Printed posters that show pictures of the farm plots and some written information about the farm hang from the sides of my booth. I will also keep a tablet running all day, playing a slide show. The more stuff you have to look at, the more you will keep people at your booth. That's important because crowds draw more crowds. Whenever you have a group of people at your stand, people passing by will wonder what's going on, and they'll flock over to take a look.

## Restaurants

Working with chefs is what allowed me to double my production from one year to the next. The first year I started I worked with only a few restaurants, as it was difficult to keep up to their demands on a weekly basis. This is why I advise working only with one or two when you're starting out, and smaller restaurants as well. You need to be careful to not take on too much at the start, because you're learning the basics of maintaining a steady weekly production system. If you overload yourself, you'll be letting down customers; I always find that it is better to under-promise and over-deliver than the other way around. The balance between what the market demands, and what you're able to produce on a consistent basis is the equilibrium you want to strive for. What makes a great farmer for a chef is not being able to show up every now and then with some great-looking product, but being able to work with them every week, to bend and flex to their demands as they change. Chefs who have some experience with farmers are great to work with because they like being limited by the season. That's what keeps their work interesting for them, but you still want to be as consistent as you can.

From a production standpoint, the best thing about selling to restaurants is that you can produce a lot more of the same thing and sell it all in one shot. For example, let's look at one crop. I can sell up to 200 pounds of radishes per week. In my farmers markets, I wouldn't be able to sell that much of one particular crop. I'm lucky to sell ten bunches at a farmers market. To sell those 200 pounds, it might just be a couple of deliveries to restaurants done in an hour or less. That's $1,000 of product very quickly.

On my farm, restaurants account for the vast majority of our sales. From a purely economic standpoint, selling to restaurants is far less work and more profit when time is factored in, because of the amount of product you can move at once, and the fact that once you deliver it, you're done and not standing at a farmers market all day. You do need to be careful because working with restaurants also brings a lot of risk. The main challenge is that if you lose a customer for whatever reason, that can deal a pretty heavy blow to your weekly income. See photo #2 (restaurants and chefs) in the photo insert section.

## Community Supported Agriculture (CSA)

CSA refers to a common form of direct consumer marketing for small farmers where customers become members of the farm. Most commonly, members pay up front for a season's worth of vegetables so that they are sharing in the risks and rewards of the farm. Every week for a predetermined amount of weeks, each member receives a basket of seasonal vegetables from the farm. Many farmers today are operating CSAs in a customized form, where members can choose from a list of veggies from the farm. New web-based software has made this a lot easier for farmers.

The CSA has become a primary market stream for a lot of small- to larger-scale farmers because it offers stability and upfront income. If you want to offer a CSA, you must have a good variety of products to engage customers; this is why CSA is not ideal for an urban farm on less than a ½ acre of land. Growing crops like broccoli and winter squash are not economical on that size of land base.

Running a successful CSA depends largely on how educated your customer demographic is about CSAs. From what I've seen, in larger cities where local food is a lot more trendy and in demand you can sell shares for top dollar, and they will sell out fast. You don't have to offer any customization, and can have few to no options for box size and pickup time—definitely an ideal scenario for any farmer. In small to medium cities, it's often a totally different story. Though a lot of people may shop at farmers markets, people don't even know what a CSA is. If you have to educate your customers, signing them up can be a lot more difficult and time-consuming. In these cases, offering more flexibility for box size, customization, pickup times and payment options will get you way more sign-ups early on. The year we changed our CSA to offer this kind of flexibility, we received more than triple the sign-ups from one year to the next. There are a few software programs out there that allow for e-commerce shopping for your members; these programs can make things really easy, but some take up to 2% of your sales. We actually used Google forms that linked to a spreadsheet. Every week we published a list of what was available, and our customers would order it through the Google form link we'd send them through e-mail.

There are four things you can offer in a CSA that will help engage a lot more customers, especially if they are new to the concept:

**1. Payment Options**

Instead of people paying for a season's worth of veggies and having to pick up one box every week, give them ability to prepay a certain amount of credit. Each time they order, the purchase amount comes off their credit total. This requires more administration, but there are software programs that can help manage this. People no longer have to pay a large sum like $500 up front; they can pay $100 every couple of months, or however much they use the service. We found that we engaged a wider base of socioeconomic levels and ages, and this plan even allowed for vacationers to sign up. We also offered an incentive to pay more up front by giving a discount based on how much you could pay. If you paid $250 or more at once, you'd get 10% added to your credit. So, if you paid $500 up front, you'd get $550 in credit.

**2. Full Customization**

Once your customers have a credit with the farm, they can order however much they want whenever they want. As long as they have a credit, they could technically spend it all at once if they desired. In order to do this, you (as the farmer) have to be comfortable enough with your production systems that you can handle a lot of potential fluctuation in weekly demand. If your members don't order for a few weeks, it doesn't matter. They are only charged when they order. You can set caps as well; we used a minimum order of $15, and you could also set a maximum if you desired.

**3. A Referral Program for Existing Members**

If a member refers someone new to sign up, give them a 10% credit bonus. So, if they signed up for a $500 box, give them a $50 bonus if they sign someone up. You can set this figure or percentage however you like. I find it's good to have credits high enough to create a lot of incentive for someone to get more people signed up. The cool thing is that, technically under this scheme, if one member signed up 10 additional people, they'd essentially get a $500 CSA share credit for free! That's a lot of incentive for a customer to help you find more customers.

**4. Multiple Pickup Times**

We used to have boxes picked up only on Fridays between 4 PM and 8 PM. The next year, we set up a second cooler on an honor system, and people could essentially pick up their box whenever they wanted. We gave a time frame so that people wouldn't be showing up at 1 AM on a Tuesday night, but the honor system cooler offered a lot more convenience for people—and that makes the CSA more attractive. We live in a small enough community that this system worked fine. However, if you were worried about theft, you could install a combination lock on the cooler and give the code to all your members.

## Small Farm Broker

Brokering product is a great way for rural and urban farmers to work together. Access to markets is often a challenge for rural farmers because of distance. While trying to engage a wider customer base, lack of crop diversity for a really small urban farm can also present a challenge. There could be mutual benefit for both types of farmers to work together. The rural farmer could be guaranteed to sell some extra product and have the ease of dropping it off once a week to one location, and the urban farmer could expand their customer reach by offering a wider variety of products to sell.

For the urban farmer in this type of arrangement, the benefit of brokering product from other growers is that she or he now has more varied offerings to engage more customers; more high-value products can be sold in each transaction. With this approach an urban farmer on ¼ acre could run a CSA program for 100 members. This would be worth about $50,000 in up-front capital, but near half of that money would be going to pay for products sourced by other growers. Still, even if half of it was for the urban farmer, that's $25,000 in up-front capital, which is really nice for any farmer. What I love about a CSA program is that the up-front money is almost like an insurance policy, because it gives you a nice foundation to cover any mishaps during the season. The only extra work would be to assemble the boxes and coordinating what products to buy from other growers, as well as picking up the products if needed. It would be reasonable to add about 25% markup to any products that were brokered, so the profit margin on $25,000 of product would be $6,250. Not that much, but consider that if you now have 100 CSA members, you are certainly selling more of the products you are growing, like salad greens, greenhouse tomatoes and microgreens. That's really where the extra value is. You wouldn't be able to offer a balanced CSA box otherwise.

Brokering for restaurants would also be a possibility here, especially if you have a good number of chefs buying from you on a weekly basis. You could be catering to smaller restaurants when brokering. For some rural farmers, based on the time it takes to assemble and deliver small orders, they don't see value in delivering to a restaurant if the order is anything less than $200. You, the urban farmer, could list some items from other growers on your fresh sheet (see page 52), and if you're delivering to those clients anyway, carrying some product from another grower would hardly be an extra cost. You'd have to decide what minimum amount (in time, money or both) would make this service worthwhile. The beauty here is that you do it on an order-by-order basis. Don't order the extra products from the farmers until you get the orders from your chefs; that way there's no risk of purchasing products that you can't sell.

It would be possible to broker product for a farmers market if you were selling at the kind of market that would allow you to do that. My farmers market doesn't allow that, so I've never tried. I would only recommend this only if you saw a huge market demand for a lot more variety than you meet yourself. The major risk for you here is that if you don't sell the product you broker, then the cost of the product comes out of your pocket.

### Retail and Other Distributors

Marketing through retailers isn't a usual approach for small farmers because you will end up making a lower price for your product. However, with the advent of some new retail business models, it is possible to sell through these outlets and still make a good profit. There are two market outlets I'm going to cover: green grocer and organic produce delivery services that serve either residential or commercial outlets.

When you look at what some of the new health-food-oriented grocery outlets are charging, it's not that hard to imagine selling your product to them at about a 35% discount and still making a profit. This is really logical only if you can move a lot of product and have the production capacity to handle high volume demand. When you're dealing with these customers, you're going to be selling by the case lot or by the hundreds of units. This could allow you to scale up the production of some of your most lucrative crops. Sell them the products that require the least amount of land, have the highest turnover and highest price. For example, if you're selling the retail business something that you'd normally sell for $3 for 25% off, that's now a $2.25 item. For a crop like tomatoes that have a long period before you harvest, that's pretty substantial considering the extra amount of work that is involved like pruning and tying them, and the fact that they occupy a larger area of the farm. However, crops like greens, radishes and particularly microgreens that occupy less area, take less work, grow faster and command a higher price are far more valuable for you to sell at discount.

I have sold through three distributor businesses before. One is an organic home delivery service, essentially like a CSA that sources through multiple farmers and has a farther reach and customer base that would be possible for my farm to establish.

Through this customer I offer a few different items, and typically they are my highest value crops. The other is an organic restaurant service that delivers produce to areas where there is little to no access to locally grown product. This is a great business model because they are servicing a high-demand niche market that is willing to pay a premium price. For this customer, I'm able to offer a wider variety, because they can pay a better price, more in line to my standard market value. The third I've sold through was a traditional health-food-oriented grocery store or green grocer. In this case, I found it worthwhile only if they were buying large volumes, and ideally for us, it was better to sell them bulk product and have them portion it accordingly, otherwise the 25%–35% discount price made it hard for us to make a profit.

### Buying Clubs

Buying clubs are usually self-organized groups of people who are passionate about canning and processing their own preserves, and they come together to buy large volumes of produce from farmers at a discount. I've most often dealt with people looking for canning tomatoes, pickling cucumbers and basil for pesto. I find these market streams valuable for an urban farmer only if you are producing more than you can sell in your usual market streams. For my farm, buying clubs are more or less a last resort because I try not to have too much of anything that would cause me to lower my prices to the point where I just want to unload it. However, there have been times where I've got a lot of basil in the field that needs to be harvested before a frost. Times like these, it'll all be harvested at once and then sold at discount, as it might have all gone to waste if it had sustained a frost.

Some community groups will organize their own buying clubs to buy from farmers. They might advertise the types of crops they're looking for on their website or perhaps a social media group. It's good to be aware of these groups, so when there is a time where you'd like to move a volume of product at once, you know where to sell it.

# Working with Chefs

The local food scene has become very trendy in recent years, and this is great thing for the small farmer. Some of the most popular shows on television and online are food related, and some chefs are considered to be A-list celebrities. Most of these chefs endorse and encourage the use of high-quality and local product. When a customer goes to a high-end restaurant, they assume that the establishment is using local product, in most cases featuring seasonal items on the menu. If you are operating your farm in an area where you have access to good restaurants like this, selling to them is an absolute must!

## Approaching Chefs

When you are looking for restaurants to work with, do some research in your area. How many restaurants are advertising the use of local produce or seasonal menus? Make a list of all the places that are offering those things. From there, do further research to see how big they are (how many seats there are), what kind of food they serve, where their price points are, where they are located, their operating times and who the chef is. You could use an app like Yelp to do some of this basic searching. Once you have a list of possible restaurants that you'd be interested in supplying, make some phone calls to those restaurants and ask if they use local or organic product and where they buy it from. In kind of a sneaky way, you could just pretend to be an overzealous customer.

The best restaurants to start with are the small ones, and as you get more experience, go bigger. After you've established a comprehensive list of potential customers to approach, it's time to drop in to talk to the chef. It's generally better to call ahead to arrange a time, but that is sometimes hard to do, as chefs are very busy people. Sometimes you just have to go in person to make the connection. Go in whenever it is not the lunch or dinner rush. If they're a breakfast place, go in the afternoon; if they're busy for lunch, go in after the lunch rush. Never go into a restaurant in the evening to talk business. Nobody does that. If you're in the restaurant as a customer, and it happens naturally, then that's something else.

When you drop in, bring a business card, a fresh sheet and some sample product, maybe just a $10–$20 sample of what you grow. Only bring in the items that you are producing consistently and in abundance. Don't bother bringing samples of things that are in short supply, because the chef may want a lot of that item. You don't want to be short when you're establishing a new relationship. Once you establish a working relationship with a chef, you can start to offer more items and one-off, special things.

In higher end restaurants, the person in charge of ordering is usually called the Executive Sous-chef or the Chef de Cuisine; sometimes it's the Executive Chef as well. In smaller establishments there is usually only one chef, and they are also often the owner. These are the people you need to talk to when dropping in. It's important to be friendly and brief. Tell the chef what you do and what you grow, and try not to take too much of their time. Ask them about what kind of products they're interested in, and if they have a local supplier already. If they do, ask if they're happy with the quality and cost of the products they are getting. If there are some things they're not finding anywhere else, perhaps there's a niche to fill. Ask what's on their menu during each of the seasons, and if you can get a sample of their menu, and past menus, all the better. This information can often be found on restaurant websites as well.

I have always found it beneficial to be up front and honest about everything with chefs. When I was first starting out, I had a small handful of chefs that I worked with, and I told all of them right off the bat that I was just starting out, and that I'd always do my best to meet their needs. Be careful not to make too many promises or guarantees when you are starting. You don't want to make promises that you're not sure you can keep.

At the beginning of each week, I send out a fresh sheet to all of my chefs. A fresh sheet is a list of all the products I offer, prices and the volume of the packages. At the top of the sheet, there are usually a few updates about what's happening on the farm or any new products available that week. I find it best to make this form mobile friendly, as many chefs like to use their smart phones for ordering, and will often text their orders.

### *Menus for the Season*

I have found that chefs are most often very interested in the seasonality of vegetables: when certain products are abundant or scarce. For the restaurants that are very focused on using local, this is the framework they use to design their menus. Every winter, I meet up with some of my best restaurant customers to discuss what they're looking to do the following season, and they're looking to hear about what kind of new things I might try too. Some of my customers, such as wineries or caterers, operate on a very seasonal basis, and they plan their year-round events such as weddings, festivals and special occasions. During the planning phase of my season, I'm marking these days on my calendar to use in my production plan. For the most part, I have found that chefs want and are willing to work with what grows best in the season, and they will always have lots of questions about what that looks like throughout the year. In cities where there is a lot of tourism, the high season for a restaurant usually falls in line with the height of the summer. And, in this period, you can expect to sell a lot more volume, so it's good to be prepared.

#### *Staying Up To Date*

If restaurants become one of your main market streams, it's a good idea to stay up to date with food trends. You could do this by following chefs on social media, reading food magazines and blogs and constantly talking to your chefs about what's coming up. One thing I have learned is that food trends always change, and you need to be constantly looking ahead to see what may be coming down the line. Often, smaller cities mimic what happens in the big cities. Places like New York, Toronto and LA are the main trendsetting cities I watch. I find that a city like Vancouver will often pick up what's happening in those places, then a town like mine, Kelowna, will often follow what's happening in Vancouver. There seems to be a trickle-down effect when it comes to trends.

#### *Record Keeping*

Record keeping is very important when it comes to working with chefs. You want to have a record of what they ordered each week, and what they received. These are often different, because you might have been short on some items they ordered. This is common as food services approach their high season, and you need to record this information so that next season you can be better prepared to meet their demand as often as possible. That just means more sales for you in the end. I always mark the common times of peak demand on my calendar, so that I can have as much as possible during that period. Many farmers that sell to restaurants will be short during this time, and if you are prepared, it's a good time to pick up more customers by coming in with product when others are in short supply.

### Specialized Growing for One-Off Events

There are times when I will plan to grow something special for a chef and event that happens just once. I call these *one-off events.* Growing to order is not something I offer to just anyone. I do this only with chefs that I trust and have worked with for a long period of time. The danger in committing to these kind of arrangements is that, if you grow something that you normally wouldn't be able to sell and that chef doesn't follow up, you're left holding the bag. So caution must be used in these circumstances. In my region the kind of events I'm referring to are field to fork, wine festivals and specific catering events. Over the years, I have learned to just have certain things ready for these events, as I usually sell more of some items that I know certain chefs will want for them. For example, we have an event here called Dîner En Blanc, where everyone wears white and all the food is white. I always sell a lot of white and light colored vegetables during that time, so I plan accordingly for this.

There are times when a high-profile chef will take a gig catering a special fundraiser event and have a vision for something very special. I've grown items like mini carrots, radishes or turnips for these types of events. In this case, they're selling tickets that can go for hundreds of dollars per plate to raise funds for something. So the cost of the product is barely a consideration. I have grown baby veggies that I would not be able to sell anywhere else, so I charge for them by the piece. The chef would come to me and say that he has 200 plates, and they're looking for 1,000 pieces of tiny carrots, 400 pieces of mini bok choy, 1,000 pieces of tiny easter egg radish; it could be a variety of things. So here I would charge anywhere from 10 cents to 20 cents per piece. When you get an order for pieces, it makes it relatively easy for planning once you know what your crops will yield.

When I'm planning one-off events during the winter, I'm looking to get an very accurate idea of what they are going to need, and I *add* that to my garden's weekly production plan. You want to make sure that these kind of events are icing on the cake for you; because they can be risky, you want to make sure you're insuring your regular weekly farm income is separate from that risk.

# Labor

When we are talking about urban farming in the way I have described here, you, as the business owner, are the primary labor source, and this is what keeps your overhead costs low, especially at the beginning. For the most part, you are managing all the day-to-day tasks on the farm from planting, harvesting and bookkeeping to delivering and building your business and brand. However, finding some assistance with labor is inevitable for all farmers at some point. Whether it's help with packing for market, pulling weeds, delivering or harvesting, you are going to need some help at some point. My hope is that, through this information, you can find better ways to streamline many of these tasks and find ways to reduce the need for labor so that you can leverage your productive capacity to its fullest potential.

## Friends, Family and Neighbors

On a very small land base, the total amount of work that needs to be done on a weekly basis is very small compared to farms at even an acre; so in this case, it can be difficult to actually make a job for someone. Another great aspect of farming in the city is that it attracts a lot of people to the idea of helping on the farm. There is mutual benefit here when you consider the value that you as a farmer can provide for some people living in the city. For city dwellers, getting to work outside provides a form of exercise, education or getting some veggies on trade.

I can get a lot of very part-time help on my farm from people in my closest network—friends, family and neighbors. In exchange for produce from the farm, people like this can come and help with simple tasks like bagging greens or portioning veggies for a few hours a week. It's a great value for both sides, and everyone wins in the arrangement.

On Fridays, I have one or two friends in the neighborhood who come over at midday to help assemble items for the farmers market. It's rarely more than three hours of work, and it's in a fun environment where they can chat or listen to music or podcasts; and the work itself is simple enough that they can go mostly unmonitored for most of that period. When the jobs are done, we just do a value exchange for veggies. Value can be set to whatever you think that time is worth. The key is that both parties need to be happy.

The other time I have used trades like this has been for one-off tasks like getting a bunch of transplants out in the spring or large harvests of crops like garlic. Since I don't do much of those any longer, there hasn't been much need for it on my end, but these trades work well for tasks like these.

I have a fantastic arrangement with my neighbors who are retired and avid gardeners. They live next door to one of my main farm sites, and I give them access to anything from the farm, year-round, and to some special areas on my garden sites to grow whatever they like in exchange for help at that one site plus the maintenance of all my tomatoes. They take care of all the pruning and tying, as well as the harvesting of all tomatoes. The neighbors take a huge task off my hands, and it is quite specialized. I couldn't give this task to just anyone because it would require a lot of training. Because they are avid gardeners that I have known for a few years, it works well. I have found that arrangements like these will pop up over time, once you have built a good amount of social equity in an area.

## Apprenticeships

Once your farm is up and running, or perhaps even before, you may have people who will approach you asking about an apprenticeship. In my experience, these arrangements operate long-term, often an entire season or most of it. You have to be careful with apprenticeships because labor laws in certain jurisdictions could make problems for you. Some farmers have abused

**FIGURE 3:** Tasks Volunteers Can Cover. Tasks are ranked in order of importance and difficulty. Bagging for market is the first and simplest task a volunteer would learn, and washing greens only when they had more experience.

| Task | Average Time/Week | Description |
|---|---|---|
| Market portioning | 3h | Filling small bags of greens or microgreens and bunching kale |
| Washing harvest bins | 1h | Washing dirty harvest bins at the washing station |
| Compost work | 30m | Dumping microgreen flats and emptying spoilage into compost |
| Restaurant portioning | 1h | Filling large case lot bags of greens and filling boxes for bulk orders |
| Pruning tomatoes | 3h | Pruning sucker branches from tomatoes twice a week |
| Picking pattypans | 2h | Picking pattypans when they are small |
| Washing root veggies | 1h | Washing radishes, beets, carrots, turnips and spring onions |
| Washing greens | 1h | Rinsing, spinning, drying and sorting salad greens |

the concept to basically just get free labor. I don't condone this at all. My advice is that if you're going to offer an apprenticeship, get everything in writing to make it clear what that person will be learning and the value exchange between the two of you. If you can offer a valuable education, then maybe you can even charge for it. I've seen some farmers take a deposit from the apprentice, and if they stay the entire season, they get that deposit back at the end; but if they leave you high and dry in the middle, you keep it. Again, you have to be careful about these arrangements, and make sure you do your homework on legalities in your area before you proceed. There is such a lack of good education in new agriculture today that apprenticeships are really the best way to learn how to farm. To create the best experience for both parties, you as the farmer, must have a clear understanding of the value you are offering the apprentice. If you're just looking for free labor, it will not work out in the long term.

## Employees

Finding good workers is a challenge for any business. Not only is it hard to find people who will work hard, but it's also difficult to find people who have a good attitude and will work passionately. In my experience it's been easy to find good attitudes and passion but sometimes a challenge to find the work ethic. Because urban farming has become trendy in recent years, it attracts people who are very passionate. I have encountered a lot of idealism (much like mine was when I was starting out) with a not so realistic perspective on the real work that needs to be done on a farm (some of the less glamorous things). When it comes to finding the right people, I have found that individuals who have a good work ethic at their foundation seem to work out the best on my farm. Having some passion about farming is important too, but only if you can actually get the work done. At the end of the day, you need to be able to make money from having employees, otherwise what's the point? If your farm is nonprofit, that's a whole other thing.

I have found it's critical to keep labor costs on my farm to less than 25% of total gross income. Ideally, I target 20%, but in the beginning, your costs may be as high as 35% of gross. Ultimately, how you calculate costs is up to you, but I have found that if labor costs get too high, there's often not enough at the end of the year to keep your bills paid throughout the off-season and then have enough capital to start up again next year. Always watch your spending, and make sure you are getting paid at least what your employees are.

### *The Importance of Systems*

When it comes to teaching people how to farm, having systems in place and consistency in how things are done is very important. This way, if problems arise, if everyone is doing the same thing the same way then you're not trying to figure out who's doing what: you're going straight to the problem itself. Whether a packaged product turns out poorly, crops germinate badly or there are problems with irrigation, if everyone is doing a particular task the same way (that you know works), then right away you can look for external reasons, like weather or other factors. During times when things need to change, find the best way to do it and make sure everyone follows suit. I always tell people that if they see a better way to do something, "Let me know, because I want to learn as well." In the past I've had some workers do one task a particular way, different than everyone else, and that has always been a recipe for disaster when the unexpected happens. Keep all tasks consistent when managing other people.

### *Building a Team*

When it comes to finding more people to grow your operation, you need to first ask yourself: what is my goal? Are you hiring people to make things easier for yourself? Because if that's the case, maybe you've taken on too much. Maybe scaling back and refining your systems would be a better and more cost-effective than hiring employees. Are you hiring people to grow the business because you need to meet a demand that isn't being filled and you see a lot of long-term potential there? I'd say that would be the best reason to hire more people. For the most part, a farm under ½ acre can be managed with two full-time people, either two partners, or one owner and one full-time person, with maybe a few part-time people for a few hours a week. A lot of this depends on the lifestyle you're looking for as well. With urban farming, running a business from a distance will not yield huge returns because the nature of the farm is small. It is possible to scale up with a lot of employees, but your margins will definitely slide down.

### *Setting Production Targets*

How much you should pay someone can come down to a pretty simple calculation based on productivity. The easiest way to quantify an employee's productivity is to have them only perform tasks related to harvesting and processing first. Once they know how to do something, see how much they can harvest in an hour, and that'll give you a good framework to calculate their wages. The best way to start this is to come up with a benchmark for every task on the farm for yourself. There can be a lot of variability here, because I don't know how well each person will work, but I do know that I work very hard and fast because the business is mine, and nobody else will work with the same incentive as I will. That's common for any business owner. If an employee can perform at 75% of the level that I can, then that's good. In some rare cases I've seen workers do the same, and on even more rare occasions, even better. If that happens, those employees are keepers, and you need to make sure they are paid fairly for their work. On my farm, I'm willing to pay people exactly what they are worth. The more value they bring to me, the more I'll pay. The sky is the limit in this regard.

For example, if I can harvest one bed (75 bunches) of radishes in an hour, that task is worth $187.50 gross profit (radishes sell at $2.50 per bunch). I would train a person to do this task as fast as I can, but I will only expect them to do 75% of that (at least at first). If an employee can harvest 56 bunches ($140 worth) in an hour, that's good value. Paying an employee $15 an hour, it took two hours of labor ($30) to produce 56 bunches packed, which sold for $140; that means your gross profit is $110 on that particular task. That's a 78% gross profit margin. When employing people, there are still other costs associated, such as government taxation, workers compensation insurance and other input costs such as fuel, and materials. Always quantify the metrics of everything you are paying for. Paying for labor is worth it only if it allows you to do tasks that cannot be delegated as easily.

# Software and Organization

One major way I stay organized and efficient on my farm is by using spreadsheets, and record-keeping software. For production, I use spreadsheets for everything: tracking yields, plantings, weekly sales, orders and crop information references. For invoicing and accounts receivable, I use accounting software. This lets me know who owes me money and makes preparing invoices fast. I also use my smart phone to make voice memos and take pictures of my plots to monitor their progress.

## Spreadsheets and Recording Data

Spreadsheets are just as important as any tool on my farm. I use spreadsheets and my smart phone more than any other tool, even more than my rototiller. Not only do I use spreadsheets in my office for a variety of applications, but I also use them on my smart phone to monitor progress on all my farm sites. With spreadsheets, it's important to collect data, but it's just as important to know how to use it, when to record it and when to apply it. I use ten forms of

### Plot Map

The logistical data on these sheets regarding where things are planted all correlate to a map I keep of all my plots. I make this map in SketchUp, and it's a simple layout of each farm plot location, how many beds there are and how they are configured. Each bed and segment is numerically categorized, and all that information is connected to locational data on the spreadsheets.

I make laminated copies of this map and keep it in all the vehicles for the farm, including the bikes. This way, if I tell someone to go and harvest a particular plot and bed, I can reference the map so that they know exactly what I'm talking about. Since the farm has multiple locations, this is very important.

data collection in spreadsheets; these are all separate pages in my software:

1. Plantings (nursery, field, microgreens)
2. Yields
3. Crop Profiles
4. Weekly Orders (present, past and harvest tallies)
5. Weekly Sales Totals
6. Land Allocation Data
7. Budget and Expenses
8. Seed Order and Stock
9. Plot Progress (dynamic sheet)
10. Spoilage

### *Sheet #1: Plantings*

On this sheet, I track everything that is started in the nursery and planted in the field and what microgreen flats I have seeded. It's important to track when things are started but also when they emerge as well as when they are mature. Over time, this data will help you better understand the date to maturity of each crop at each time of the season. This sheet is directly related to the Yields sheet, and some of the first few columns (such as Date, Crop and Location) are exactly the same; that way you've got some correlation that makes things easy to find once you sort the table by any of those categories.

### *Sheet #2: Yields*

With the yields sheet, we're tracking everything that comes off the field: where it was harvested, when and how much. This is how you monitor whether a crop or plot is actually making you money, and the data helps you get a better understanding of what you can expect in the future. All of the averages of expected harvest on my crop

**FIGURE 4:** Plantings: Sample Sheet.

| Date (m/d) | Crop | Location (plot/sg#/ bed#) | Crop Variety | Bed Size | NS, PT, DS, TR | DOE (m/d) | DTH (m/d) | DTM | Bed Rows | Plug # | Seed Vol. | Notes |
|---|---|---|---|---|---|---|---|---|---|---|---|---|
| 03/11 | arugula | law/tun1/04a | bellezia | 23 | ds | 03/18 | 04/21 | 41d | 9 | | 12 | |
| 03/11 | arugula | law/tun1/04b | silvetta | 23 | ds | 03/18 | 04/21 | 41d | 9 | | 15 | |
| 03/11 | radishes | law/tun1/03 | fb | 46 | ds | 03/17 | 04/13 | 33d | 7 | | | |
| 03/11 | red Russian | law/tun1/05 | | 46 | ds | 03/18 | 04/09 | 29d | 9 | | | |
| 03/12 | mizuna | law/tun2/02b | red frills | 23 | ds | 03/18 | | | 9 | | 35 | |
| 03/12 | tatsoi | law/tun2/02a | | 23 | ds | 03/18 | | | 9 | | 35 | |
| 03/13 | Hakurei | lawr/tun2/01 | | 34 | ds | 03/23 | | | 7 | | 7.5 | |
| 03/27 | kale | garths/3/01 | | 30 | tr | | | | | 108 | | |
| 03/27 | kale | garths/3/02 | | 30 | tr | | | | | 108 | | |

*Note*: Date (M/D): date of planting; Crop: name of the crop; Location (plot/sg#/bed#): plot name, segment number and bed number; Crop Variety: for example, a certain type of radish; Bed Size: the length of the bed; NS, PT, DS, TR: how planted: nursery, potted up, direct seeded or transplanted; DOE (M/D): date of emergence (when the crop emerged from the soil); DTH (M/D): date to harvest (when the crop was harvested); DTM: days to maturity calculated from Date and DTH; Bed Rows: how many rows were planted; Plug #: how many transplants were planted; Seed Vol: how much seed was used; Notes: special notes if needed profiles sheet come from years of accumulating data from these yields sheets. The better you track this information, the better you can plan in the future. If you know what your demand will be for any particular crop and you know what you can expect for a yield, allocating land and planning your farm becomes very simple. Also, by tracking all of your yields, you can easily compile the total sum for your entire harvest for the season to see exactly how much you pulled off the field.

### *Sheet #3: Crop Profiles*

This is a reference for all the crops that I grow. It's a collection of years of information, and I use it to set standards for the way planting and harvesting are done. I keep a laminated copy of this sheet alongside my plot map for reference in all my vehicles, bikes and truck. Feel free to add more information from Part 10 on a sheet like this for easier reference.

### *Sheet #4: Weekly Orders*

Weekly orders are compiled and sorted using three spreadsheets: orders past (see Figure 7), orders for the week (preorders) (see Figure 8) and the harvest tally sheet (see Figure 9). I sort the preorder sheet in four ways, then copy and paste each sorting into a text document and print each one; I have up to four separate sheets on a clipboard:

1. All restaurant orders, farmers market items and landowner shares and trades *sorted by item*
2. Restaurant orders *sorted by item*
3. Restaurant orders *sorted by customer*
4. Farmers market items, landowner shares and trades *sorted by customer*

**FIGURE 5.** Yields: Sample Sheet.

| Date Harvested | Crop | Location (plot/sg#/bed#) | Yield lbs or bun | Area/ Flats Hrv | Yield Ratio (fl/ft) | Profit/ crop/ flat | Crop # | Date Planted | DTH |
|---|---|---|---|---|---|---|---|---|---|
| 2015/03/29 | spinach | law/tun1/1 | 57 | 34.5 | 1.65 | 342.00 | 1 | 2014/10/01 | 179d |
| 2015/03/29 | spinach | law/tun1/2 | 57 | 34.5 | 1.65 | 342.00 | 1 | 2014/10/01 | 179d |
| 2015/03/29 | spinach | law/tun2/1 | 41 | 46 | 0.89 | 246.00 | 1 | 2014/10/01 | 179d |
| 2015/04/02 | mustard | law/tun2/2a | 2.5 | 23 | 0.11 | 25.00 | 1 | 2015/03/06 | 27d |
| 2015/04/02 | red Russian | law/tun1/5 | 4.75 | 23 | 0.21 | 47.50 | 1 | 2015/03/06 | 27d |
| 2015/04/10 | mustard | law/tun2/2a | 5.75 | 10 | 0.58 | 57.50 | 1 | 2015/03/06 | 35d |
| 2015/04/10 | red Russian | law/tun1/5 | 9 | 23 | 0.39 | 90.00 | 2 | 2015/03/06 | 35d |
| 2015/04/15 | spinach | law/tun2/1 | 39 | 46 | 0.85 | 234.00 | 2 | 2015/03/06 | 40d |

NOTE: Date Harvested: date you harvested; Crop: name of the crop; Location: where the crop was harvested; Yield (lbs or bun): how many pounds or bunches were harvested; Area/Flats Hrv: how many feet of a bed were harvested, or how many microgreen flats were cut; Yield Ratio (fl/ft): yield ÷ area = yield ratio; Profit/crop/flat: yield × price of crop = profit; Crop #: How many times bed was harvested; Date Planted: when crop was planted; DTH (days to harvest): date harvested – date planted = DTH.

**FIGURE 6.** Crop Profiles: Sample Sheet.

| Crop | CVR 5/5 | Available for harvest | Avg DTM from seed date | DS/ TR | When to DS | When to TRN | Avg Yield/ 25' bed | Avg Yield/ Cut | Avg Yield/ Ft. | # of cuts/ bed | Profit per 25' bed | Price per lb or unit |
|---|---|---|---|---|---|---|---|---|---|---|---|---|
| Arugula | 5 | spring | 35 | ds | Mar–Oct | — | 30 | 12 | 1.2 | 2–3 | $300 | $10.00 |
| Basil | 3 | summer | 70 | tr/ds | July–Aug | Apr–Jun | 25 | — | 1 | — | $250 | $10.00 |
| Beet greens | 4 | summer | 40 | ds | May–Aug | — | 30 | 10 | 1.2 | 4–6 | $240 | $8.00 |
| Beets | 4 | spring | 72 | ds/tr | June–Aug | Apr–May | 100 | — | 4 | — | $300 | $3.00 |
| Bok choy | 3 | spring | 50 | ds/tr | June | Apr–May, Aug–Sep | 50 | — | 2 | — | $250 | $5.00 |
| Carrots (baby) | 4 | summer | 65 | ds | Apr–Aug | — | 60 | — | 2.4 | — | $240 | $4.00 |
| Carrots (full size) | 4 | summer | 78 | ds | Apr–Aug | — | 75 | — | 3 | — | $225 | $3.00 |
| Cilantro | 4 | spring | 30 | ds | Apr–Sep | — | 250 | — | 10 | 1–3 | $500 | $2.00 |
| Dill | 4 | spring | 60 | ds | Apr–Aug | — | 200 | — | 8 | 2–3 | $400 | $2.00 |
| Salad turnips | 5 | spring | 38 | ds | Apr–Sep | — | 100 | — | 4 | — | $300 | $3.00 |
| Kale | 4 | early spring | 80 | tr | — | Apr/Aug | 115 | — | 4.6 | — | $575 | $5.00 |
| Lettuce | 5 | spring | 45 | ds | Mar–Oct | — | 30 | 10 | 1.2 | 2–4 | $270 | $9.00 |
| Lettuce heads | 5 | spring | 45 | tr | Mar–Oct | — | 75 | 25 | 3 | 3–4 | $675 | $9.00 |
| Mustard | 4 | spring | 35 | ds | Mar–Sep | — | 30 | 10 | 1.2 | 2–4 | $270 | $9.00 |
| Parsley | 4 | summer | 70 | ds/tr | June–Aug | — | 235 | — | 9.4 | 3–4 | $470 | $2.00 |
| Pea shoots (flat) | 4 | early spring | 14 | ds | Any | — | 0.85 | — | 0.14 | — | $13 | $15.00 |
| Radishes | 5 | spring | 28 | ds | Mar–Sep | — | 75 | — | 3.00 | — | $188 | $2.50 |
| Radish shoots (flat) | 4 | early spring | 12 | ds | Any | — | 1 | — | 0.17 | — | $20 | $20.00 |
| Red Russian kale | 5 | early spring | 30 | ds | Apr–Sep | — | 40 | 8 | 1.60 | 2–8 | $320 | $8.00 |
| Scallions | 4 | spring | 70 | ds/tr | May–Aug | Apr | 40 | — | 1.60 | — | $320 | $8.00 |
| Spinach | 5 | early spring | 45 | ds | Mar–Oct | — | 35 | 15 | 1.40 | 1–3 | $245 | $7.00 |
| Summer squash | 3 | summer | 60 | tr | — | May | 80 | — | 3.20 | — | $320 | $4.00 |
| Sun shoots (flat) | 4 | early spring | 12 | ds | Any | — | 1 | — | 0.17 | — | $15 | $15.00 |
| Swiss chard | 3 | spring | 65 | tr | — | Apr | 65 | — | 2.60 | — | $325 | $5.00 |
| Tatsoi | 4 | spring | 35 | ds | Mar–May, Sep–Oct | — | 30 | 10 | 1.20 | 2–4 | $270 | $9.00 |
| Tomatoes (Cherry) | 3 | summer | 145 | tr | — | May | 180 | — | 7.20 | — | $720 | $4.00 |
| Tomatoes (Heirloom) | 3 | summer | 165 | tr | — | May | 330 | — | 13.20 | — | $825 | $2.5 |

Note: Crop: name; CVR: numerical crop value rating; Available for harvest: season crop is available to harvest; Average DTM from seed: average time crop takes to mature from seed; DS/TR: How the crop is most commonly planted (direct seeded, transplanted or both); When to DS: months to direct seed this crop; When to TR: months to transplant this crop; Avg Yield/25' bed: average yield from a 30" × 25' bed; Avg Yield/Cut: average yield per harvest (Cut and Come Again Greens); Avg Yield/ft: average yield per linear foot of bed, useful if you want to calculate what to expect from beds that aren't 25' long; # of cuts/bed: (for Cut and Come Again Greens) average amount of cuts per bed; Profit/25' bed: gross profit from one 30" × 25' bed; Price per lb or unit: the average price per pound or bunch unit.

**FIGURE 7.** Orders Past. This sheet can be very long, as it will include all of your past orders for the season.

| Date | Customer | Items | Units | Unit Wgt (lbs) | Total Weight (lbs) | Price | Value | Total Sold |
|---|---|---|---|---|---|---|---|---|
| Monday | Distributor order | Arugula (wild) | 8 | 2 | 16 | $20.00 | $160.00 | |
| Monday | Distributor order | Salad turnips (12 ct) | 8 | 12 | 96 | $40.00 | $320.00 | |
| Monday | Distributor order | Kale | 1 | 5 | 5 | $25.00 | $25.00 | |
| Monday | Distributor order | Pea shoots | 1 | 2 | 2 | $30.00 | $30.00 | |
| Monday | Distributor order | Radishes EE | 20 | 1 | 20 | $2.50 | $50.00 | |
| Monday | Distributor order | Radishes FB | 10 | 1 | 10 | $2.50 | $25.00 | |
| Monday | Distributor order | Red Russian | 5 | 2 | 10 | $20.00 | $100.00 | |
| Monday | Distributor order | Spring Mix (mild) | 7 | 2 | 14 | $17.00 | $119.00 | |
| Monday | Distributor order | Sun shoots | 1 | 2 | 2 | $30.00 | $30.00 | |
| Monday | Distributor order | **Total** | | | | | | $859.00 |
| Tuesday | Winery restaurant 1 | Arugula (wild) | 2 | 2 | 4 | $20.00 | $40.00 | |
| Tuesday | Winery restaurant 1 | Beets golden | 20 | 1 | 20 | $4.25 | $85.00 | |
| Tuesday | Winery restaurant 1 | Beets red | 20 | 1 | 20 | $4.00 | $80.00 | |
| Tuesday | Winery restaurant 1 | Carrots (baby) | 16 | 1 | 16 | $4.00 | $64.00 | |
| Tuesday | Winery restaurant 1 | Salad turnips (12 ct) | 2 | 12 | 24 | $40.00 | $80.00 | |
| Tuesday | Winery restaurant 1 | Radishes EE/FB | 15 | 1 | 15 | $2.50 | $37.50 | |
| Tuesday | Winery restaurant 1 | Tomatoes (cherry mix) | 10 | 1 | 10 | $4.00 | $40.00 | |
| Tuesday | Winery restaurant 1 | **Total** | | | | | | $426.50 |
| Tuesday | Small cafe | Arugula (wild) | 1 | 2 | 2 | $20.00 | $20.00 | |
| Tuesday | Small cafe | Spring mix | 3 | 2 | 6 | $17.00 | $51.00 | |
| Tuesday | Small cafe | **Total** | | | | | | $71.00 |

From sorting #1 I add up the amounts for each crop, and create the current week's Harvest Tally Sheet. When we are in the field harvesting, we don't need to know what each order is, just how much of each crop we need to harvest in total.

Using sorting #2, we assemble restaurant orders one crop at a time. For example, we'll allocate all the radishes into each order box for restaurants that ordered radishes, then move on to the next product.

I use sorting #3 to double-check that each order is correct. Sometimes, I'll make changes to this sheet if we are short on particular items for any orders. Then, I'll use this sheet again right before I head out to deliver. I never print invoices until all orders are assembled, just in case changes are made based on lack of any items.

I use sorting #4 to portion produce in the same way. Items for the farmers market are often in four-ounce bags, and so are

**FIGURE 8.** Orders for the Week (Preorders). This sheet can be sorted in four different ways for packing and harvesting purposes. Here it is sorted by crop.

| Date | Customer | Items | Units | Unit Wgt (lbs) | Total Weight (lbs) | Price | Value | Total Sold |
|---|---|---|---|---|---|---|---|---|
| Friday | Breakfast/Lunch Rest. | Arugula | 1 | 2 | 2 | $20.00 | $20.00 | |
| Friday | Cocktail Bar | Arugula | 2 | 2 | 4 | $20.00 | $40.00 | |
| Friday | Medium-Sized Rest. 2 | Carrots (baby) | 8 | 1 | 8 | $4.00 | $32.00 | |
| Friday | Medium-Sized Rest. 3 | Carrots (baby) | 20 | 1 | 20 | $4.00 | $80.00 | |
| Friday | Medium-Sized Rest. 3 | Cilantro | 4 | 1 | 4 | $2.00 | $8.00 | |
| Friday | Vegan Rest. | Cilantro | 4 | 1 | 4 | $2.00 | $8.00 | |
| Friday | Medium-Sized Rest. 2 | Salad turnips (6# case) | 0.5 | 6 | 3 | $40.00 | $20.00 | |
| Friday | Medium-Sized Rest. 3 | Salad turnips (6# case) | 0.5 | 6 | 3 | $40.00 | $20.00 | |
| Friday | Medium-Sized Rest. 1 | Salad turnips (6# case) | 0.5 | 6 | 3 | $40.00 | $20.00 | |
| Friday | Medium-Sized Rest. 3 | Pattypans | 10 | 1 | 10 | $4.00 | $40.00 | |
| Friday | Medium-Sized Rest. 1 | Pea shoots | 2 | 2 | 4 | $30.00 | $60.00 | |
| Friday | Medium-Sized Rest. 2 | Radishes EE/FB | 4 | 0.5 | 2 | $2.50 | $10.00 | |
| Friday | Cocktail Bar | Radish shoots | 1 | 1 | 1 | $20.00 | $20.00 | |
| Friday | Medium-Sized Rest. 2 | Radish shoots | 1 | 1 | 1 | $20.00 | $20.00 | |
| Friday | Medium-Sized Rest. 2 | Red Russian | 9.5 | 2 | 19 | $10.00 | $95.00 | |
| Friday | Medium-Sized Rest. 3 | Red Russian | 1.25 | 2 | 3 | $20.00 | $25.00 | |
| Friday | Breakfast/Lunch Rest. | Spring mix | 4 | 2 | 8 | $17.00 | $68.00 | |
| Friday | Medium-Sized Rest. 3 | Spring mix | 1 | 2 | 2 | $17.00 | $17.00 | |
| Friday | Vegan Rest. | Spring mix (box) | 1 | 10 | 10 | $85.00 | $85.00 | |
| Friday | Breakfast/Lunch Rest. | Spring onions | 1 | 1 | 1 | $8.00 | $8.00 | |
| Friday | Cocktail Bar | Sun shoots | 1 | 1 | 1 | $15.00 | $15.00 | |
| Friday | Medium Size Rest. 1 | Tomatoes (cherry mix) | 10 | 1 | 10 | $4.00 | $40.00 | |
| Friday | Medium Size Rest. 2 | Tomatoes (cherry mix) | 10 | 1 | 10 | $4.00 | $40.00 | |
| Friday | Medium Size Rest. 3 | Tomatoes (cherry mix) | 20 | 1 | 20 | $4.00 | $80.00 | |
| Friday | Vegan Rest. | Tomatoes (cherry mix) | 10 | 1 | 10 | $4.00 | $40.00 | |
| Friday | Breakfast/Lunch Rest. | Tomatoes (heirloom) | 15 | 1 | 15 | $2.50 | $37.50 | |
| Friday | Medium Size Rest. 3 | Tomatoes (heirloom) | 13 | 1 | 13 | $2.50 | $32.50 | |
| Friday | Medium Size Rest. 1 | Zucchini (baby) | 1.5 | 1 | 2 | $4.00 | $6.00 | |
| Friday | Medium Size Rest. 3 | Zucchini (baby) | 3.5 | 1 | 4 | $4.00 | $14.00 | |
| Friday | Breakfast/Lunch Rest. | **Total** | | | | | | **$133.50** |
| Friday | Cocktail Bar | **Total** | | | | | | **$75.00** |
| Friday | Medium Size Rest. 1 | **Total** | | | | | | **$126.00** |
| Friday | Medium Size Rest. 2 | **Total** | | | | | | **$217.00** |
| Friday | Medium Size Rest. 3 | **Total** | | | | | | **$316.50** |
| Friday | Vegan Rest. | **Total** | | | | | | **$133.00** |

**FIGURE 9:** Harvest Tally. This sheet is taken out to the field for harvesting.

| Crop | Total Order | In Stock | Lbs Harvest | Crop Location(s) | Notes |
|---|---|---|---|---|---|
| Arugula | 6 | | 0 | Bernard/02/06 | crop the whole bed |
| Carrots | 28 | 11 | 6 | Wash/01/15 | |
| Cilantro | 8 | | 17 | Wash/03/08 | |
| Salad turnips | 9 | 1 | 8 | Wash/01/09 | |
| Pattypans | 10 | | 8 | home | |
| Pea shoots | 4 | | 10 | home | |
| Radishes | 2 | | 4 | Bowes/01/07 | |
| Radish shoots | 2 | | 2 | home | |
| Red Russian | 21.5 | | 2 | Bowes/02/04&05 | crop both beds |
| Spring mix | 20 | 5 | 22 | Bernard/01/03 | crop the whole bed |
| Spring onions | 1 | | 15 | Bernard/02/13 | |
| Sun shoots | 1 | | 1 | home | |
| Tomatoes (cherry mix) | 40 | 15 | 1 | home | |
| Tomatoes (heirloom) | 28 | 7 | 25 | home | |
| Zucchini (baby) | 5 | | 21 | home | |

the items that are going to landowners and trades.

### *Sheet #5: Weekly Sales Totals*

This sheet compiles the total sales each week for the entire season. I stack total weekly sales from every previous year into this sheet as well. This way, I can compare what I've sold each week to what I did in previous years. This window into the past and future allows me to quickly and easily look back and see what I've done in past seasons. On it, I highlight weeks of high-volume sales and also make notes of when I was short on any particular items. After a couple of years of data, this sheet allows me to spot trends and to predict weeks of high sales. This data has been critical in allowing me to navigate the changing market much more effectively. (Find this sheet at theurbanfarmer.co.)

### *Sheet #6: Land Allocation Data*

This reference sheet contains all the logistical data regarding my farm. Each plot and segment is listed, with the dimensions of the plot, number of beds, length of beds and the plot type (whether it's Hi-Rotation or Bi-Rotation).

### *Sheet #7: Budget and Expenses*

I use this sheet to plan my budget for the season: how much I will spend on certain purchases as well as new capital investments I will make. I use a separate sheet to track expenses. Every couple of months, I'll

**FIGURE 10.** Land Allocation Data.

| Plot | HR/ BR | Beds # | Bed Size(s) | Total Linear Bed Feet | Total Bed Square Footage | Plot Dimensions | Plot Square Footage | Total Plot Area | Total Beds | Total Linear Bed Feet | Total Beds in 25’ Units | Total Plot Square Footage | Farming Since |
|---|---|---|---|---|---|---|---|---|---|---|---|---|---|
| Home Base front yard | HR | 5 | 35 | 175 | 438 | | | | | | | | |
| | HR | 1 | 24 | 24 | 60 | | | | | | | | |
| | HR | 1 | 10 | 10 | 25 | | | | | | | | |
| | HR | | | | | 30' × 35' | | 1,050 | 7 | 209 | 8 | 523 | 2012 |
| Home Base tunnel 1 | BR | 6 | 46 | 276 | 690 | 18' × 46' | 828 | | | | | | |
| Home Base back 4 beds | BR | 4 | 46 | 184 | 460 | 6' × 25' | 150 | | | | | | |
| **Total** | | | | | | 49' × 46' | | 978 | 10 | 460 | 18 | 1,150 | 2012 |
| Washington Ave main segment | HR | 12 | 50 | 600 | 1500 | 50' × 50' | 2500 | | | | | | |
| Washington Ave tunnel 1 | BR | 4 | 36 | 144 | 360 | 12' × 36' | 432 | | | | | | |
| **Total** | | | | | | | | 2,932 | 16 | 744 | 30 | 2,932 | 2014 |
| Bowes Ave sg 1 | BR | 6 | 25 | 150 | 375 | 16.5' × 25' | 412.5 | | | | | | |
| Bowes Ave sg 2 | BR | 7 | 25 | 175 | 438 | 21.5' × 25' | 537.5 | | | | | | |
| Bowes Ave sg 3 | BR | 5 | 30 | 150 | 375 | 15' × 30' | 450 | | | | | | |
| **Total** | | | | | | | | 1,600 | 18 | 475 | 19 | 1,400 | 2010 |
| Bernard Ave sg 1 | HR | 16 | 25 | 400 | 1000 | 67' × 31' | 2077 | | | | | | |
| Bernard Ave sg 2 | HR | 7 | 23 | 161 | 403 | 25.5' × 25' | 637.5 | | | | | | |
| **Total** | | | | | | | | 2,700 | 23 | 561 | 22 | 2,715 | 2010 |
| Smith Ave main plot | HR | 11 | 53 | 583 | 1458 | 36' × 54' | 1944 | | | | | | |
| Smith Ave small segment | HR | 3 | 33 | 99 | 248 | 15' × 31' | 465 | | | | | | |
| **Total** | | | | | | | | 2,500 | 14 | 682 | 27 | 2,409 | 2015 |
| | | | | | | | | 11,760 | 88 | 3131 | 124 | 11,129 | |
| Acres | | | | | | | | 0.270 | | | | | |

just copy all my receipts into this, so that at tax time there is less information gathering to do. There are two examples of the budget: Operating and Start-Up. (Find these sheets at theurbanfarmer.co.)

### *Sheet #8: Seed Order and Stock*

There are two sheets here. Every winter before I complete my seed order, I take stock of all the seed I still have. Generally I try to order new seed each year, but because I buy high-quality seed, I can usually get at least two years of good germination rates. After I take stock, I subtract what I have from what I need, then I order the difference. (Find these sheets at theurbanfarmer.co.)

### *Sheet #9: Plot Progress*

This sheet is critical when your farm is multilocational. I don't have all of my crops in one place, so I can't easily walk outside and see what will be available for harvest in the coming week. But knowing what's available is critical when I speak with chefs and take orders. Once per week, I record an assessment of each plot into the voice recorder on my smart phone; I describe the progress of each bed that is near maturity. Each bed for every plot is listed on it in numerical order, and this sheet basically acts like a still photograph displaying what's in each bed and when I can expect it to be ready. It helps me communicate with my customers what I have coming, track when beds will be turned over and rotated, as well as communicates to employees about what needs

**FIGURE 11.** Plot Progress.

| Location (plot/sg#/ bed#) | Bed Size | HR/ BR | Current Crop | Date Planted | Next Crop | Crop 1 | Date Planted | Crop 2 | Date Planted | Crop 3 | Date Planted | Crop 4 | Date Planted |
|---|---|---|---|---|---|---|---|---|---|---|---|---|---|
| Washington Ave main segment | | | | | | | | | | | | | |
| WA/01/01 | 50 | HR | OW carrots | 08/03 | | radishes | 04/01 | mini heads | 05/13 | OW carrots | 08/03 | | |
| WA/01/02 | 50 | HR | OW carrots | 08/03 | | radishes | 04/01 | mini heads | 05/13 | OW carrots | 08/03 | | |
| WA/01/03 | 50 | HR | OW spinach | 10/01 | | spinach | 04/01 | salad turnips | 06/05 | mini heads | 07/13 | OW spinach | 10/01 |
| WA/01/04 | 50 | HR | OW spinach | 10/01 | | mini heads | 04/07 | radishes | 06/30 | arugula | 08/03 | | |
| WA/01/05 | 50 | HR | OW spinach | 10/01 | | red Russian | 04/01 | beets | 07/08 | OW spinach | 10/01 | | |
| WA/01/06 | 50 | HR | OW spinach | 10/01 | | arugula | 04/10 | radishes | 06/05 | mini heads | 07/08 | OW spinach | 10/01 |

to be done. This sheet is directly correlated to my laminated plot map.

#### *Sheet #10: Spoilage*

Everything that doesn't sell gets recorded in this spoilage sheet. This information is valuable for a couple of reasons:

1. It gives you a clear visual for what you are not selling. You can use that data to decide if you are growing too much of something or it's not worth growing at all.
2. It can be useful for tax purposes. Spoilage is considered an expense in some tax jurisdictions, but you need to be careful here.

Each country, state or province will have different regulations about the percentage of that spoilage you can claim against your income. Do research to see if you can claim spoilage in your area. (Find this sheet at theurbanfarmer.co.)

### Using Voice Memos

I prefer to use a voice memo app on my smart phone to record data, mainly because, when I'm in the field my hands are dirty, and I don't want to stop and tinker around with my phone. I find it much easier and faster to use a voice memo and just say what I did or whatever I need to do, and input the data later. After I plant a series of beds, I'll just speak what I did into the recorder; I do the same for harvesting or keeping track of the progress of each plot. I schedule two times a week to input this data into my spreadsheets. Usually, I'll do this when I'm sitting at my desk in the morning over coffee; on Tuesday morning, I'll input what was planted, and on Friday morning, I'll input what was harvested. With the advent of recent voice recognition software on smart phones, it is now possible to record this information straight into a text document that is synced to a document on your computer.

### Invoicing and Accounts

All of the accounts of my restaurant, distributor and wholesale clients are kept in an accounting software. These programs can be purchased for anywhere from $100–$600. If you are dealing with a lot of accounts they are worth every penny. When an invoice isn't settled when it's delivered or if payment is due later, you can easily lose track of who owes you money. I find that spreadsheets don't make payment tracking easy. Accounting software can be used for printing invoices as well. It takes some time to set up, because you must create profiles for all your customers and input all the products you sell. Once that's done, however, the program makes fast work of printing invoices, because each customer and item will autofill. So, with just a few keyboard strokes, I can put together an invoice that has multiple items and information on it.

For managing accounts, I really like accounting software because I can sort my invoices as paid or open. Open invoices are ones that haven't been paid yet. By sorting it this way, I can easily see who owes me money and how much, and can quickly prepare a statement of what they owe and e-mail it to the customer instantly. Sometimes chefs get behind on paying, so you need to stay on top of this. I purchased my program for $600, and there are lot of choices out there. Most accounting software does the same things.

# Self-Promotion

Promoting yourself and your farm is a critical part of achieving success. One thing that will work in your favor all the time is that urban farming is considered good news, and it's becoming incredibly trendy. Urban farming is still so unique that it's seen as a social movement by most. This gives urban farmers a huge leg up when promoting themselves. People will want to hear your story, and you'd better prepare to tell it over and over again. The best testing ground for telling your story will be farmers markets, but you'll eventually be contacted by other forms of media, whether it's bloggers, local newspapers, radio or TV stations. Never pass up these interviews when you're getting started. The more you can do early on, the more of a foundation you'll lay out for the future. Being the first person in your town to start urban farming will also give you a significant advantage in the future. This is called *first mover advantage,* and it's why I encourage new urban farmers to start in a place where nobody else is doing it. If you can establish yourself as the first mover, you'll most likely always maintain a solid share of the market in your area.

## You Are Your Own Brand

Since direct consumer marketing is the most beneficial form of selling for most small farmers, you as the farmer become the face of the business. Thus, in a sense, you are your own brand. Your customers will want to get to know you and will want to hear your story. I have heard so many interesting stories over the years about how and why people decided to get into farming. Those stories are always interesting, inspiring and different from person to person.

In my experience, the part that seems to get the best reactions from listeners is *why.* Why do you do what you do, and what led you down the path to start farming? These are the questions you want to be prepared to answer when you're promoting your farm, and ultimately, yourself. The what and how are important as well, but when you are first introducing yourself to outlets in the media, the *why* is what will engage your listeners. Learn to craft your story in brief talking points, and then get to the what and how shortly after.

When you are approaching chefs or those kinds of customers, starting with what you do and offer, might be better because you want to first appeal to their immediate interests to justify why you are there taking up their time. But, once you break through the initial barrier, everyone will want to know your personal story and why you do what you do.

### Media

Before I even started farming, I had people in the media contacting me to see what I was doing. Here's another advantage of starting in a place where nobody else is urban farming. When I started developing my initial farm site in late summer of 2009, I had people stopping by constantly. At that time, I wasn't even technically farming, but people in the area were so curious about what I was doing that it drew a lot of attention. Because of the compost program I started that fall, I was spending five to ten hours a week picking up compost from my friends, and this started to get peoples' attention. I had CBC radio, local newspapers and bloggers all calling me for interviews. I took every opportunity to give interviews. Looking back, starting that compost program gave me something to talk about early on in the development of my farm. It was a good foot in the door early on with media outlets.

### Customer Relationships

The most important aspect of promoting yourself is building strong customer relationships. Remember Pareto's Principle, the rule of 80/20? Focus on the 20% of your customers who bring you 80% of your business. Make a concerted effort to remember peoples' names, especially your 80/20. If you can't remember names, write them down on a list. I used to have a note of all of my consistent customers on my smart phone. I'd make notes about their appearance, then put their names down.

The key is to build strong relationships with your customers, and they will sing your praise. Remember, a satisfied customer is your best salesperson. When you as the farmer are the primary brand, you need to maintain a good and positive public image of yourself.

# Finance Options

When you're starting your farm, you'll need to spend some money to get things started. I spent only $7,000 for initial infrastructure my first year, which wasn't that much. However, I spent a considerable amount of time finding deals on my main purchases, which in the end saved me a lot of money up front. I would advise everyone to do the same. You should be able to get everything you need for a small urban farm on ¼ acre for around $10,000 or less. If you don't have this money saved, then you're going to have to look into finance options. There are a number of options out there, especially in unconventional areas.

Since urban farming is still such a new idea in the conventional business world, the traditional avenues of financing are generally slow to come around. The biggest challenge when trying to find funding is convincing people that farming on small land bases can be profitable. There are just not many mainstream examples out there. I always promote the path of least resistance with everything, and looking for funding is no exception. First, go to friends and family, look at crowdfunding and save the traditional avenues like banks and government as a last resort.

### Family and Friends

Go to your family and friends first. Since the investment required is nothing that substantial, borrowing from a few friends or family members might be a simple way to assemble the cash you need. If you're only looking for around $5,000, you might be able to find that among a few different people as well as family members.

### Community Bonds

Community bonds funnel investment into local sustainable infrastructure such as farms, community social projects and land restoration projects. For a farmer, creating a bond can be looked at like requesting a CSA type of investment. A number of people would invest a certain amount in the farm, receive interest dividend payments in the form of cash or product and receive all of their money back after a defined period. The basic idea here is to get some investors to put in some larger sums of capital (like $1,000 or more). For five years, you will give them a 5% dividend each year in either product from your farm or cash. (For example, if an investor were to give you $5,000, for five years you would give them $250 in cash or product per year—that's the interest payment—then at the end of five years, they get their $5,000 back.) So, participating in a community bond is a way of making their money work slowly to improve local food resilience, but also to get a small return for their investment. The benefit to the investor is that, instead of their money sitting in a bank and collecting a tiny amount of interest, that money can collect the same interest but go to work and help grow new farmers. It's an investment in the future of farming.

### Crowdfunding

With the advent of web-based technology, it is possible to raise money in a totally new and decentralized way. Crowdfunding has been the sole source of funding for a lot of new start-ups, even urban farms.[1] If you can make a compelling story through video and creative online marketing, it's possible to raise hundreds of thousands of dollars. With this platform, you need to set a target funding goal. If you don't achieve it within your time frame, you won't see any of the money; it's all returned to your investors. So, for a new farm it might be best to start with a smaller figure (like maybe $5,000 or so). But, if you have the ability to make a really compelling campaign and know how to leverage your traffic online, the possibilities are endless. If you're in need of some start-up capital and are a little web savvy, this may be the best place to look for funding because you don't have to jump through all the hoops of traditional funding: applications, credit checks or taking on debt.

### Credit Unions and Banks

Going through traditional banks and credit unions can be challenging for urban farmers, particularly if you don't own land or have a traditional farm lease. Some of the new credit unions are starting to get used to these different applications, but for the most part, you'll probably get a strange look when you ask for funding for a farm that is based on front and backyards in the city.

I have seen some credit unions on the west coast make incredible progress in funding small farms. They are looking more at the farmer's character than just their financial reports, which is critical for helping new farmers.

As far as the big traditional banks go,

save these for the last resort. What you will often hear is that they want you to generate income all year. That can be a challenge for a lot of farmers, though of course it's possible. If you're looking for funding from a bank, make sure you have a solid year-round plan that shows steady income 12 months of the year. A lot of banks don't like to lend to any seasonal businesses because of the risks involved.

PART 4

# FINDING THE RIGHT SITE

Finding the right place in your city to farm will be your first and most important decision. Make sure you have read Chapter 4 (The Zones of Your Farm and Your Life) before you read this part.

# 15 Scouting for Land

In this part, I describe the best way to actually go about looking for land once you have decided on the area where you'd like to be based. There are a variety of tactics we can employ when looking for land. Some will work better than others, some involve using technology and some are as simple as leveraging your friend and family networks.

There are some materials that you should get together before you start your search:

1. Write up some material about who you are, why you want to farm and what your farm is all about.
2. Print these up into some catchy flyers that you can post around on bulletin boards at places like health food stores, community centers and garden clubs.
3. Make up some business cards for yourself that you can pass around when you meet people. Carry business cards with you at all times, in case you meet someone and need to have them remember you. Always make sure to get their name and information as well so you can follow up.
4. Build yourself a simple website that describes what you want to do, so you can direct people there.

Make sure to get the word out to as many people as you can. Talk to bloggers and media, and use social media as often as you can so that everyone you already know is made aware of what you're planning to do.

### Mapping Software

Before you start pounding the pavement in search of urban plots to farm, do some scanning of your desired area with mapping software. Use mapping software like Google Earth or Maps to search your area. Your city also might include a mapping tool on its website. I have used this in the past, and it often will have more detail about lot size, roads and easements than Google will.

Using software at the start can also help you find a desired part of town more easily. If you find that some areas have larger lots than others, then you could decide to start looking in those places first. Even today when I receive land offers, once I have vetted the landowner to see if they are the kind of person I'd like to work with, I'll put their address into mapping software to look at the size and aspect. This online research could save a trip by finding out some basic things first.

### Low-Hanging Fruit

The first place to look is for the low-hanging fruit. Ask all of your friends and family for some land you can start with. Maybe it's your grandmother's front yard or something like that. The key is to get something started as soon as you can, and that will become a pivot point to other opportunities. Starting somewhere, getting something going and letting people find you is the best way to find land!

### Door to Door

If you are going to knock on doors, prepare yourself for some rude reactions. Many people will assume, if you're knocking on their door, that you're an evangelist of some kind; they are often not ready to hear what you have to offer. I have found it better in the past to put a flyer in the mailbox and not bother people at the door. Let them come to you.

### A Logistical Land Checklist

When I am looking at a new plot of land, there are ten factors that land should offer. Some of these factors I can be flexible with, and some I will not. However, any prospective plot must offer a good number of factors before I agree to farm there.

One thing that is important to remember when you're looking at plots to farm is that farmers are scarce. Today, only 2% of people in North America are farmers, but there is land everywhere. Often, the people that own it don't know how to use it for farming. This means that you as a farmer are very valuable, and there are many options for you. Approach all negotiations from a place of abundance. You are the one in demand, because you are scarce. Land is abundant! Don't ever feel like you have to take a piece of land that isn't ideal or if it seems the landowner might be difficult to work with.

#### Land Checklist

1. The landowner
2. The site history
3. Soil test
4. Size and location
5. Invasive weed check
6. Available light at all seasons, shade obstructions
7. Fencing
8. Visibility to the public
9. Water access
10. Accessibility (Time and Entrance)

### *1. The Landowner*

Before I go and look at a site, I will first have a quick phone call with the prospective landowner. There are somethings I'm listening for that could disqualify this plot right away. First, what is the landowner's general demeanor? Are they easy to talk to, friendly, easygoing? How did they hear about you, and why are they interested in having you farm their land? These are good questions to start, and the answers will reveal whether they're going to be the kind of person you want to work with for years to come.

Once I ask them a few questions, I then briefly explain to them how my program works. I'll first outline what they get out of the deal—$20–$25 of veggies per week from May until October, a beautiful looking garden—and how they don't have to do any of the work, and we cover all the costs for setup. I like to start with what's in it for them. In most cases they already know, but early on in your career, you'll have to make a few more pitches, and people like to hear about what they get first. Once you've walked them through that, you'll get a sense really quickly about what they think.

It's important to ask landowners what their needs and concerns are. I have had some landowners ask about the amount of vegetables they get, and they think they should get more, based on the size of their land, or they'll want me to take on things like mowing their lawn for them or pruning their shrubs. It's good to be very clear about what you're offering, so there is no assumption that you're there to offer a landscaping service. In some cases with really ideal and large pieces of land, those things can be worked out, and sometimes special services may be worth it for you to be able to farm there. With my first plot, that was my home base for four years, I agreed with the owners to look after the general maintenance of the property. Based on what I was getting from there, it was well worth it at the time. Basically, you're just trying to get a sense of the person to see if they're the kind of person you could see yourself working with.

### *2. The Site History*

Before you even look at the site, you want to learn about its history. Ask how old the house is, and what the plot in question has been used for in the past. Was it a garden at some point? Were there ever any other structures built there? If so, what were they? The main thing you're trying to figure out here is whether the site has potential soil contamination from heavy metals, industrial chemicals or glyphosate use. In dense urban areas, if you're looking at a site that has been used for a gas station or some kind of other industrial use, there's a good chance that there may be heavy metals such as lead or mercury in the soil. If it was part of a conventional farm or orchard then you may well find traces of glyphosate in the soil. With most suburban backyards that I've encountered, there has been no history of industrial use. But, if you're concerned about this, then you will need to test your soil.

Next door neighbors on all sides of the property can also be a concern. If someone next door is spraying herbicides, watch for that as well.

### *3. Soil Test*

If I'm worried that a site is contaminated, I won't use it at all. Especially if there are plenty of other sites to look at. If the site is otherwise very ideal but there may be a concern about contamination, it's up to you if that is going to be worth it. For one, soil tests for heavy metals can be expensive; secondly, if the soil is contaminated, then you're going to need to go through a remediation process or just bring in new soil. I would advise against this, because it will drastically increase your start-up cost. However, if you have a situation where new soil can be provided and/or you're willing to make the up-front cost, then by all means, do what works for you.

Besides contamination, our main concern is with soil depth, organic matter, macronutrients and pH. Nutrients, macronutrients and pH can all be adjusted fairly easily. Soil depth however, can be a little more complicated and can take more resources. Some urban sites have a very shallow soil profile: when you stick a shovel in the ground, you'll see just a few inches of topsoil, and then a very hard-packed heavy mineral, straight sand clay or gravel. I try to avoid soils like this, but they can be brought into fertility with the addition of new soil and good compost. From what I've seen over the years, most soils can be brought up to fertility with some time, effort and capital. Read Chapter 21 on plot preparation for more on this. Testing the site for macronutrients and pH is a very easy thing to do: simply a matter of getting a small soil testing kit from the garden center and taking it from there. These soil testing kits come with instructions on how to do the tests. It's a very simple process, easy to figure out.

### *4. Size and Location*

The size and location of a plot is very important, and what area might be available is one of the first things I ask about when I'm talking to a prospective landowner. My minimum requirement for plot size is 2,000 square feet. I'll look at smaller plots only if they're really close to my main site. For example, if my next door neighbor or someone else on my street offered me there plot and it was 1,000 square feet, I would consider it. If the site is small but close, it can become one of your Hi-Rotation sites and still be economical.

As far as location goes, this is something that will improve over time. When you first start out, you will probably have to take some plots that are not quite as close together as you'd like. There has been a natural progression inward of my plots over the years. I have cycled around 20 farm plots over five years, and they get closer and closer together every year. This is because you will build a reputation in the areas you farm, and the more people that see what you're doing, the more offers of land you'll get. My first year, all of my sites spanned three miles; now they span a mile and a quarter.

### 5. *Invasive Weed Check*

It's very important to know the main invasive weeds that could be detrimental to farming in your bioregion. If you see a lot of any particular invasive weeds, especially those that are rhizomatous, you probably want to avoid the site all together. My main show stoppers are Field Bindweed (*Convolvulus arvensis*) and Canada Thistle (*Cirsium arvense*); I won't use a site if I see a lot of either of those. Annual weeds such as dandelion, mallow, purslane and chickweed can all be mitigated over time. See photo #3 of invasive field bindweed in the photo insert section.

### 6. *Available Light at All Seasons—Shade Obstructions*

If you want to farm a particular plot for the longest possible season, then you're going to need to have a lot of light. Learn where the sun is at every time of the day and in every season in your local area. Once you understand the movement of the sun, you can walk onto any plot and see where the light will be at any time of the day and any time of the year. This is very important, because if there are big obstructions like fences, trees or buildings, they could block the light that you need for growing. Most food plants need at least six hours of light per day to grow, and some more than others. Any summer crops like tomatoes or peppers need lots of light, especially at the hottest time of the day (at least for us in the cooler climates of North America). However, sometimes shade at certain times of the day can work to your advantage when you grow cooler weather crops. I have a few plots where there is a building on the west side, so that in the summer during the late afternoon, the plots fall into shade. That's a great thing because it gives the crops a break from heat at the hottest time of the year. Shade also gives me a break from the heat if I need to work there around that time.

The most important thing you're looking for is open aspect to the south (in the Northern Hemisphere). If there are any tall buildings or fences directly to the south, that's going to cause a lot of light deficiency during the shoulder seasons when you need it most. A little bit of shade in the morning or afternoon is manageable for the most part. See photo #4 (shady plot) in the photo insert section.

### 7. *Fencing*

Finding a site with a fence isn't totally necessary, but it is a good thing. Some neighborhoods are better fenced than others, and you'll have to do your own assessment to see if not having a fence is going to be an issue. Fences will stop people, deer, dogs and other large creatures from cutting through your garden or helping themselves to your veggies. It's possible to install some inexpensive temporary fencing, like a simple deer fence, but that doesn't look as nice. Many times in the past a landowner was willing to pay for a fence to be built, if it meant that I would farm their site.

#### 8. *Visibility to the Public*

How visible your farm plot is is very important; it can work as its own form of marketing for you. In the past I've taken on a site just because it was in a nice neighborhood with good visibility. One of the great aspects of being an urban farmer is access to community. When people see you out there working hard, and the beauty of the gardens themselves, they get very excited. Be prepared to hear—on a daily basis—how lovely your garden is and how folks love what you're doing. I have met many good customers over the years simply because they walked by and saw me working. Plot placement is a huge marketing advantage that is often overlooked. I've used plots that didn't meet all my criteria for size and location simply because they were a good showcase. I call these Flagship plots. See photo #5 (a publicly visible plot) in the photo insert section.

#### 9. *Water Access*

Because it's potable and clear of contaminants in most cases, having access to municipal water is one of the greatest advantages of urban farming. And water is usually conveniently located near the house, so it's just a matter of tapping into. When you're scouting a site, you want to make sure that you do have access to this water. Sometimes, if the house is far away from your plot or the landlord wants access to water as well, you will have to install a main line feed from the house to a manifold close to your plot. I will usually install some kind of manifold to split from the spigot, then run a main line and build a small water main close to the site.

Providing access to water is the responsibility of the landowner, and if they can't provide water, that is a major deterrent from farming there. Water usage on all of my plots has been small enough that my landowners don't notice any difference in their bills, except for the two hottest months, July and August. In these months, I usually compensate them for the increase in their water bill. That has been around $100 per plot for those two months. Water prices are different everywhere, so there are no real benchmarks. It's up to you to decide what's economical or not, based on costs in your area.

#### 10. *Accessibility (Time and Entrance)*

Can you get access to your potential plot every day during regular business hours? Ideally, I want to be able to come in unannounced any day and any time between 7 AM and 6 PM. For the most part I take Sundays off; but sometimes you never know, you might have forgotten something and need to get in there. It's good to be clear about this when negotiating with your landowner. If they want to be notified every time you come by, that could be a bit of a nuisance to do consistently. But, if you can come to an agreement for something like Monday through Friday anytime between 7 AM and 6 PM, then make a phone call if you're going to come in on a Saturday or Sunday, that would be fair and easy enough to work around. For all of my plots, I don't run any loud machinery before 9 AM and after 5 PM. Some landowners like to be notified when I'm coming to rototill, and that's a reasonable request.

Access in and out of the plot is also very important. Is there a gate that you need to go through? How wide is it? Will you be able to fit your tiller through, as well as other pieces of equipment you'll need to move in and out on a regular basis? I have left plots in the past that were too difficult to get in and out of on a consistent basis.

# Urban, Suburban and Peri-Urban Land

There are three types of urban land: urban, suburban and peri-urban. Each has its own benefits and drawbacks.

## Urban Land

Urban land has the advantage of best location. In most cases it'll be closest to where your customers are going to be, but this factor does depend on the size of your city. In a larger city (with over 500,000 people), urban land will be in the densely populated areas, and in these cases there are logistical challenges to take into account:

- The land has a higher chance of being contaminated, so you'll need to study the history carefully.
- Plots can be very small.
- If there are large buildings nearby, they might greatly affect available sunlight.
- The land might be located in an area where crime and vandalism occur.

In large cities, multi-plot farming can be a challenge logistically, especially if you can't get a good number of plots in a concentrated area. I've seen many urban farmers with as many as 20 500-square-foot plots; this isn't very ideal and means a lot of trucking around. It is possible to be successful in this context, but close attention must be given to how your plots are laid out. You will most likely need to make small monocultures to streamline your workflow as much as possible. Without strategic planning, having 20 tiny, mixed-garden plots spread around a city can be a perfect recipe for early burnout.

The most successful farmers I have seen in dense urban areas are either farming on parking lots with raised beds or on rooftops, or they have high-tech hydroponic or aquaponic operations on rooftops that involve polycarbonate, heated greenhouses. In all of these cases, the initial start-up cost can be tremendously high. However, there can be some advantages to farming this way as well. These farms can often be high-profile because they are so unique and are such a spectacle in highly concentrated urban areas.

Other value can be stacked beyond just farming as well. Parking lots can be good places to teach gardening and farming workshops. Rooftop farms can also work in a similar way but can also become good spaces to host foodie events or even weddings. Because the rooftop farm is such an interesting landscape, it offers a lot of potential for multiple revenue streams. In the case of all three of these examples, with the amount of infrastructure cost going into the sites, a proper legal land lease is absolutely necessary if one does not own the land. You would also need to negotiate a long-term lease in order to justify up-front costs.

### *Parking Lots*

Parking lots (hardscape) with raised beds are probably the cheapest truly urban option of the three. In this case, build container-style raised beds between eighteen inches and two feet off the ground; fill them up with all new soil. Usually three- to four-foot-wide beds are good. If you were to build a ¼ acre farm with 16 beds at 4 feet wide, 18 inches high and 100 feet of total length, and if you were to spend $30 per cubic yard for new soil and/or compost, you would spend $10,656 on soil alone. You would most likely spend around $5,000 on lumber and other materials to build the beds, and if there is no fencing, that would have to be taken care of as well. Fencing could easily add another $5,000. For the rest of the infrastructure, you would probably spend another $10,000 to get all your major tools and implements. So you are looking at a $30,000 minimum investment to start a ¼ acre farm on hardscape.

### *Soil-Based Rooftops*

These urban farms are a little more common on the east side of North America where annual snowfall is a lot higher and the roofs are built to handle heavy loads. The soil used on rooftop urban farms is a special composite material that is lightweight. It includes little hollow rocks in place of natural rocks. Unfortunately this soil can be very expensive (upwards of $80 per cubic yard). If we had a ¼ acre rooftop that was 104 feet square and we brought up 8 inches of special soil at $80 per yard, soil alone would cost $21,364. Additional costs would be hauling the soil, plus a liner to separate the rooftop from the garden. Because of wind on top of buildings, a lot of crops must be staked or trellised. So, even here, at a best case scenario, you are probably looking at a minimum $50,000 start-up for a ¼ acre soil-based rooftop farm. Keep in mind that this is just cost for beds and soil, and not equipment and infrastructure.

### *Soil-Free Rooftops*

Rooftop hydroponic operations are becoming more popular in super-high-density cities such as Montreal, Singapore and Beijing. Most of the farms that are within ¼ acre employ some vertical growing strategies. These farms cost in the millions to establish and are very high risk. I believe that large corporations will be moving towards these kind of farming systems. I think overall it's a good thing and will be part of a natural migration of people moving into cities; these farms will supply a much-needed demand for fresh produce. However, the costs involved to set up such farms are out of the reach of most people.

## Suburban Land

The suburbs can offer good size, location, access, less contamination and better light for urban farm plots. Plots in suburbs will be a lot bigger in general. Based on my city's size (population 117,000), I consider the land that I farm to be suburban, even though my farm is right downtown; lots in my city look like suburban lots. This type of land offers good market access, lot size and potable water as well. Going forward I see huge potential in the suburbs for farms that will feed cities.

As more people leave the suburbs behind to move into the cities for work, suburbs could see sharp declines in real estate values; cheaper land along with affordable housing rental and purchase prices might entice more farmers to move there and start farming. It's true that, where transport costs are concerned, living in the suburbs to serve the city may not be as ideal as living in the city, but suburban location could still be a good compromise considering the start-up costs of some of the urban farm models I have mentioned. When looking at the distances between rural land and almost any city, the suburban farm is a perfect meeting place. A drive into the city once or twice a week for markets or delivering is far more manageable than driving in every day to go to work. When it comes to bringing farming to the masses, I think the suburbs offer the most potential, as suburban land is far more accessible to a wider group of people.

## Peri-Urban Land

Peri-urban areas lie between rural and suburban. These plots of land would have been small farm homesteads at some point—and in a lot of cases they still are. Peri-urban lots in my city often sit on one and two acre parcels with either giant lawns, horse farms or some other small hobby farming. This kind of land can offer a lot of the advantages that suburban land has, but with far less public visibility and reliable water access. These places are often on well water, which can make shoulder season or winter farming a challenge. One nice thing about peri-urban farming is that you could probably centralize your entire operation onto one plot of land.

# Multiple or Single-Plot Farming

## Multi-Plot Farms

Farming on multiple plots has benefits and drawbacks. The major drawback is that you have to set up separate irrigation systems; this costs a little more than if your farm were centralized. You can't as easily walk out into one field and see how everything is doing. And the biggest drawback is your travel time between sites. True that these are all drawbacks, but I think the benefits of multiple plots outweigh them:

1. If you can't secure a long-term lease on a plot of land, having a series of smaller plots works in your favor because if you lose one site, the whole farm isn't at risk.
2. The more plots you have, the more neighbors you have. This helps farmers find customers and supporters more easily.
3. A diversity in microclimates is a huge benefit. If some extreme weather event affects one plot, it might not affect the others.
4. Pests and weeds can better be avoided. Some plots are more isolated and thus sheltered from weed seeds blowing in. Pests can also be avoided by crop rotation among plots.

One year in particular, we had a freak hailstorm in mid-August that lasted three hours. Hail the size of golf balls completely pummeled our two-acre peri-urban site. We sustained $30,000 in crop damage. The lettuce looked like a string trimmer had shredded it. Meanwhile, all our small plots in the city were completely unaffected. Having that diversity was what kept my farm afloat throughout the season after that storm.

Fertility on a multi-plot farm is a little bit more complicated. Moving compost around can be laborious, so a more compact program for fertilizing maybe required. It's a lot easier to move around compost teas and extracts with the use of backpack sprayers, and using dried materials such as meals is far more practical then moving buckets of compost between each plot.

## Single-Site Farms

The biggest advantage of having a single site is being able to concentrate your infrastructure and production. Let's face it—it is a lot easier to do things on a day-to-day basis if you are in one place. You can pay less attention to organizing the workday if everything is all in front of you. I think most farmers would prefer to have a single site, and even I would find it more convenient. However, it can be difficult to find one site that meets all of your criteria, especially a site you can hold on to it for a considerable amount of time.

Obviously a single site doesn't have some of the advantages of a multi-plot farm, but there is more potential to sell from the farm gate and teach farming or gardening (if you so desire). Also, a single site can have a simpler and more convenient fertility system: you can just have one, big pile of finished compost that sits in one place all year.

# Urban Soil

**Disclaimer!** I am in no way a soil scientist and am not formally educated in this field. I describe methods below that have worked on my farm over the years. I realize that soils are different everywhere, so these methods will in no way work across the board for everyone.

Soils we often see in urban environments are marginal at best. In the case of suburban lawns, we're not often dealing with contamination but more just with soils with little or no nutrients and organic matter. In my experience, people who most often donate their lawns for urban farming are not the type to have a manicured lawn. They're usually reluctant to have a lawn in the first place, and would way rather see someone like you put it to good use. Turning a lawn into a garden can present some challenges, but nothing that isn't manageable.

## Invasive Weeds and Grasses

Most commonly, a new potential farm plot I'm looking at is covered with invasive grass: Twitch or Quackgrass (*Elytrigia repens*) or Bermuda grass (*Cynodon dactylon*). They are all basically the same thing. The grass has rhizomatous roots that grow a few inches under the topsoil, and it spreads as the roots grow underneath concrete or whatever obstruction is in the way. These plants are incredibly resilient and can break up sidewalks in the course of two years if left unchecked. When a plot of invasive, rhizomatous grass is rototilled, the tiller tines will cut up the rhizomes. This will multiply the rhizomes and can sometimes make the grass more invasive. To get rid of invasive grass like this, till it only in dry weather, and then rake out the rhizomes over a period of weeks. Don't be discouraged about invasive grass. It has never stopped me from getting a site into production. Yes, it may take a little bit longer than if it were domestic grass, but it's manageable.

Always methodically look for invasive weeds. There are many types out there, and some are worse than others. Some you can manage, and some you must avoid at all costs. It's possible to manage small isolated areas of these weeds, but in my own area, if I go to a new potential plot and see a lot of either Field Bindweed (*Convolvulus arvensis*) or Canada Thistle (*Cirsium arvense*), I will not take on the site. In the past, I have had plots with both of these weeds, and they got so out of hand that I suffered 90% crop failures as a result. This is not a good scenario for any farmer.

This happened because I didn't do the research to find out what weeds would be a problem in my bioregion before I took on those plots. Please take my advice, and find out which invasives are the worst in your area, and avoid them at all costs! When you don't own the land and might be there only a few years—and if you have other land options—it's not worth the effort. Explore all other options first.

### Contamination

Soil contamination is more common on land in densely populated areas, less frequent in suburban or peri-urban land. However, soil contamination is possible anywhere. First, mitigate contamination by researching the history of the potential site. Find out what was there before, if there has been a history of chemical use or whether the site was once a gas station or anything industrial. These sites are commonly referred to as brownfields. If I think a site is contaminated, I will just avoid it altogether as long as there are other options.

Where your only option is to use an industrial site, the best way to remediate soil is to just build up. To remove soil and replace it with new soil takes a lot of time and resources, and there's still no guarantee that plant roots will not encounter heavy metals. The one thing about heavy metals such as lead or mercury is that they are pretty immobile. They won't move up your soil; they will just sit statically. So building soil up, on top of contamination, is the most common option. In this case, build raised beds or large garden boxes.

If you are building beds right on top of contaminated soil, you will need to construct a barrier between your new soil and the bad soil. This could be something like a layer or two of heavy landscape fabric with a layer of sand on top of that, followed by between 18 inches and two feet of new soil.

If you are building over top of concrete, you don't need to worry too much about a barrier between the beds and the soil below, because the concrete is a barrier.

I am no expert in dealing with contaminated soils, and I would advise against urban farmers working with them. I have heard reports that community garden plots built on top of brownfields were finding traces of contamination even though they had laid down landscape fabric at the bottom of the beds.[1] One thing to keep in mind is that plant roots can even work their way through concrete over time, and landscape fabric alone may not be enough to stop them from growing into contaminated soil. I believe that future farming on brownfields will have to be done through non-soil-based systems like hydro or aquaponics.

I have heard of people using crops like sunflower to remediate soils as the plants will pull heavy metals out over time. This is a possibility for cleaning the site, but such remediation can takes years and a lot of effort. This isn't an option for a farmer who wants to get into production as quickly as possible.

## Hardpan

Most urban, suburban and peri-urban soils are going to have what's referred to as *hardpan*: beneath the top eight inches of soil, there is a layer of heavy compaction. After rototilling a site for the first time, it'll be obvious where that is because you'll notice that the tiller made the top eight inches fluffy and nice, but beneath that you'll notice a hard-packed surface.

Hardpan must be addressed if you are going to grow vegetables on that soil; otherwise the roots of the vegetables that hit that layer may be prevented from reaching water or nutrients. Most annual vegetables have shallow roots at first, so hardpan isn't a major issue early in site development. But once the site has been laid out, beds formed and all invasive grasses have been dealt with, paying attention to hardpan will only improve your production. To deal with hardpan, see Chapter 21: Turning a Lawn into a Farm Plot.

## Rubble to Soil

Much of the soil I have used in the past is a form of clean fill (basically just dirt, like sand and clay, with some rocks in it). There is little or no organic matter in it, and for the most part, all it's good for growing is weeds. As long as this soil is not contaminated, it can be converted into a useful growing medium by spending time to rake out rocks, bringing in large amounts of good finished compost and adding organic fertilizers. The choice of fertilizer depends on NPK or pH levels in the soil.

With some sites, I went through all the stages of plot development, and once my beds were formed and subsoil dealt with, I applied many inches of compost to just the beds, and did a shallow till to mix it in. On one particular site, we applied three inches of compost to the beds, with good amounts of a nitrogen-based organic fertilizer, and worked it in; these beds were in production weeks later, and they continue to produce over the years.

## Ongoing Fertility

On all of my urban plots, I apply at least two inches of compost to each bed every spring, and sometimes I make a mid-season application if I feel that it is needed. I also apply dried meals and/or liquid applications of compost tea and sea kelp. Compost each time would be my preferred method, but I am somewhat limited because I'm on a multi-plot farm and transport is far too time-consuming.

However, I have had a lot of success with my compact and lightweight systems. The compost teas, liquid sea kelp and dried fertilizers like blood meal or dried manures seem to keep organic activity up just fine. With a decrease in tillage during the hot months of summer, my soils have stayed alive with bugs, worms and general biological activity.

### *Application Between Crops*

Between my crop rotations in my Hi-Rotation beds, I use compact fertilizers between each crop. I use two small coffee cans' worth of a broad-spectrum organic fertilizer, which is a mix of a bunch of ingredients for one 25-foot bed. There is no formula that can apply to every farm and every soil here. This is what I use, and it has worked for me, but your farm soil may require something totally different. You must constantly be sure that your crops are looking healthy throughout the season, and doing some mid- and end-of-season fertility tests will help you know if you're on the right track. In some of new sites that had really poor soil to start, I will bring in a second application of compost mid-season to keep fertility and soil composition on track. But for the most part, I apply my compact fertilizers between each crop all season. The types of compact fertilizers that I have used are bone meal, alfalfa meal, blood meal, dried manures or a mix of some or all of the above. I have recently settled on using a dried organic turkey manure fertilizer that has an 8/2/4 NPK rating. It has worked well between crop successions in my Hi-Rotation areas. Keep in mind that Hi-Rotation crops are typically low feeders, meaning that they don't pull very much nutrient from the soil. Radishes, lettuce and spinach are low feeders. Crops like tomatoes are considered heavy feeders.

# Land Agreements and Leases

For urban farmers growing in front and backyards, the key to fast and cheap start-up is to avoid buying land. To lease or rent land, you need to understand how land lease agreements work. I am not a legal expert; what I'm sharing with you here is based entirely on my own experience. If you are concerned with actual legalities of contracts, you need to find your own legal counsel to get definitive answers.

For most of the land I farm, I use a standard Memorandum of Understanding (MOU). This doesn't have to be a legally binding contract, though it can be. The form I use outlines everything regarding each party's needs and responsibilities so that there are no gray areas. With all my garden plots, I pay the costs for use of the land in fresh weekly produce, delivered to or picked up by the landowner. Each landowner receives a fixed amount (between \$20–\$30 each week (retail value)) of produce each week for the season I farm on their land. For sites where I can extend the season, the landowner will receive veggies for however long we are farming there.

## Negotiating Terms and Base Agreements

When you've got to the point where you want to move forward with a piece of land, you will need to have a meeting with your new landowner to go over all the things you'll need from them as well as all the things you will provide. During negations you need to be very clear with your landowner on every point in your lease, memorandum or rental agreement, and have everything signed off so both parties are absolutely clear and there are zero gray areas of understanding.

## Memorandums of Understanding (MOUs)

I use a standard agreement with most of my landowners. MOUs for plots of land that aren't absolutely critical to my operation—for example, major infrastructure such as greenhouses or my base of operations equipment. This agreement is not legally binding, but it could be if both parties want it to be.

I ask all landowners for at least a three-year commitment because of the work that goes into preparing and maintaining a site; however, if they have to back out after a season, that is OK. Sometimes something may have changed in their life that they didn't foresee, and they need you to leave. In those cases, I always ask that I be able to finish the season, so that their decision is at no major cost to me. You'll find a sample MOU agreement at theurbanfarmer.co.

### Leases

A lease is what I use for all my long-term plots. If I have greenhouses or use the site as a base of operations, I want to have a legally binding lease in place. Sometimes for these plots, I pay a yearly fee for the use of the land, and do not trade for vegetables. In the case of one of my plots in the past, I paid the yearly tax on the site, which was around $2,000 per year. Not cheap as farm land is concerned, but when one Hi-Rotation plot can generate $30,000 on 3,000 square feet in one season, paying land tax can still be economical. You'll find a five-year lease I have used in the past at theurbanfarmer.co.

### Rentals

You may meet someone who owns a few rental properties and might like the idea of someone like you managing them instead of them paying someone to maintain the landscaping. In these cases you could just pay on a monthly basis, or however you'd like to arrange it. In these cases a base figure that is reasonable would be $50 per 2,000 square feet per month. In this case, you're looking at $600 per year, which is very reasonable. If you are saving that person time and money from landscaping, then you are also bringing value to the agreement. If they were willing to take veggies instead of some or all of the cash (which is more ideal for you), then you could work that into the equation. I choose to trade product as often as I can.

# Urban Pests

An urban farmer must deal with many types of pests. In my experience, the least problematic are traditional agricultural insects because I can run away from them by rotating certain crops from plot to plot. I find, more than any others, it's mammals that create more problems. The best strategy for pests is to follow the path of least resistance. First, identify where the problem is: Is the pest attacking a particular crop? Is it a problem during a particular season? If so, perhaps there is a passive solution. Instead of fighting the pest, perhaps that crop can be moved to another site, or maybe that crop shouldn't be planted during that season.

## Birds and Mammals

Good fencing should be your first line of defense for dealing with mammals such as *deer, dogs* and *humans*. If your landowner has a decent fence, these critters shouldn't be too much of an issue. *Cats, marmots* and *raccoons*, on the other hand, can get over or under any fence. I have dealt with many situations over the years, and I've tried a variety of techniques to deal with them (such as catch and release). The best and easiest way to keep *cats* off your soil is to keep it moist. If you prepare some new beds and don't plan on planting them for a couple of days or a couple of weeks, water them right away and even then cover them with tarps. You may do that as a stale seedbed technique anyway, and this will work to keep cats off. The key is to keep your soil moist after it's been planted. Once a crop is established, cats usually won't go digging around. Their favorite soil is fluffed up and dry. Trapping seems to be the only thing that works with *raccoons*. It's best to consult with your local SPCA, city, regional district or a pest specialist for these situations.

First figure out what the animals are eating or destroying at the site, then go from there. The only birds that have been a problem for me are *quail.* The best way I have found for dealing with them is to use floating row cover fabric over newly planted crops. Birds like to pick the tops of newly sprouted veggies, so if you can keep them covered until the crop is more developed, they won't be as much of a problem. But again, the same applies here too: figure out what they like best, and try planting that somewhere else if you can. For *rats, mice,* and *voles,* baiting and trapping works best; first figure out where they are coming from, and then focus on that area with a more aggressive strategy. I once had a group of rats living in my compost pile. I had never had problems with rats before, let alone ever seen one. But I noticed that the rats were eating all of the turnip beds that were close to the compost pile. I immediately stopped planting turnips at that site. This made them have to venture farther from their nest to look for food, and the local cats noticed this. I also did place traps. Soon after, I barely had any rat problems because the cats took care of them for me. The main idea here is that if the pest has to travel farther for its food, it may become vulnerable to other predators in the area—this is a systems thinking approach.

The top pest for the urban farmer is other *humans.* Sometimes people raid your garden and pick some veggies here and there. In my experience it has never been too much of any issue. For the most part, if someone were to go in and pull some carrots, cut some greens or pull some kale, I might not even notice. Just like dealing with dogs, good fencing is the best way to mitigate people raiding your garden. This is also a neighborhood issue as well. You may notice that some areas in your city are more prone to pest problems than others. The first thing is to assess the damage, see how often it's happening and then from there decide what steps can be taken to best deal with the problem.

Setting up a hidden game camera can be a great way to find out what or who the problem is. These cameras are used by hunters and cost around $80. You can hide them somewhere like in a tree or near the roof of the house. The camera will take a picture every 10 seconds, or however frequently you set it. I have found these cameras to be very useful for investigating any type of mammal pest, whether cats, dogs, raccoons or humans.

A passive technique would be to not plant any easily harvested crops (like tomatoes, peppers or roots) on that plot. I have never seen or noticed anyone come into my plots, get on their hands and knees and harvest greens with a knife. So, perhaps if you think you might have theft in your high traffic areas, grow the crop that is least likely to be stolen.

### Insects

Insects will always be a problem for farmers, and there's really no way to avoid them. Take a proactive approach by using row covers, or avoiding planting that crop during that insect's gestation cycle. For example, cabbage root maggot is always a problem for my radishes and turnips during the later spring. Since I need to keep planting these crops all season, I use 80 gram insect netting to cover all those crops during that time. The netting doesn't work very well in the heat of the summer because it's too hot under the cover and the heat makes the radishes and turnips bolt. Fortunately for me, those pests aren't as much of an issue during summer.

Another example are aphids on my kale. Aphids become a problem during the summer, but aren't an issue at any other time of the year. I have tried using nettings to protect my kale, but I find that aphids still find their way underneath it. So my approach here is to not grow kale from late July into August. I pull all of it out at the end of July; then in the third week of July, I start a new crop for fall, then transplant it at another plot around the first or second week of August. I am avoiding the problem altogether here, and not bothering to fight it.

### Defensive Location Rotation

One of the best parts about being a multi-locational urban farmer is that you can run away from a lot of problems by planting a problematic crop at another plot. Over the years, I have been able to avoid most pests becoming a systemic problem at all of my sites because of this technique. For example, every year that I plant early bok choy at a site, by the time the crop is just about mature I start to notice flea beetle moving onto the crop. I will plant only a small amount of it at one site at a time for this reason, and by the time I harvest that crop, I start to plant it at another site. Since I'm taking away that pest's main food supply, it doesn't have the opportunity to populate. From here, I'll just keep planting crops that flea beetle isn't attached to. This works very well for any crops that are in Hi-Rotation; since they have short growing cycles, the lead time it takes to react to that problem is relatively short. Location rotation is my main strategy for dealing with all pests. First get that crop off site, and then come up with a proactive technique to avoid it again so that it never gets established from the start. Because of all the physical obstructions that exist in a city (like buildings, roads, fences, shrubs) pests have a harder time following crops around. On a single site farm, if you have aphids on one end of the field, they're going to find the food they like on the other end. It's really that simple, and why being on multiple plots can be a great thing.

PART 5

# BUILDING YOUR FARM, ONE SITE AT A TIME

There are a number of elements involved when it comes to developing your farm. Whether you decide to farm on a single site, or go down the multi-plot road, a lot of the core infrastructure will be the same. But where you decide to put core infrastructure will be especially important if you have multiple locations.

Credit: Katie Huisman

**1** Farmers market.

Credit: Katie Huisman

**2** Restaurants and chefs.

**3** Invasive field bindweed dug out of the ground; notice the long root system.

**4** Notice shade from the south stretching over this plot.

**5** A publicly visible plot.

**6** A landscape fabric perimeter.

**7** 50 × 50-foot hi-rotation plot, five-zone overhead irrigation system.

**8** Main water lines run from irrigation box.

**9** Irrigation box: five zones for the HR area, the other three for other parts of the plot.

**10** Hi-rotation, four-zone overhead system, odd-shaped plot. Also note one long bed, temporarily combining two shorter segments.

**11** Micro sprinklers in a greenhouse.

**12** Drip irrigation: four lines on a thirty-inch bed.

**13** A simple small nursery (12′ × 20′), utilizing vertical space.

**14** This indoor vertical nursery can also be used for microgreens production in winter.

**15** Poly tunnel doubling as a vertical nursery.

**16** Use the poly tunnel as a temporary work area for early nursery work: here, potting up tomatoes on a soil mixing table.

**17** Small, three-cylinder import truck loaded for market.

Credit: Andrew Barton

**18** Electric-assist utility bike with two custom-built steel-frame trailers.

Credit: Katie Huisman

19 Hand harvesting spinach: bunch in the left hand and cut with the right.

Credit: Katie Huisman

20 Collect as much as possible, holding with the left hand before dumping produce into the bin.

Credit: Andrew Barton

21 Hand harvesting lettuce; see also interplanted tomatoes amongst the lettuce.

Credit: Andrew Barton

22 Harvesting tatsoi with the Quick Cut Greens Harvester; notice interplanted tomatoes to the left.

Credit: Andrew Barton

**23** Harvesting sunflower shoots: bunch with the left hand and cut with the right.

Credit: Andrew Barton

**24** Harvesting radishes: bunch with the left hand and use the right hand to bunch with an elastic band. Tomatoes are already interplanted; removing the radishes frees up space for them grow.

Credit: Katie Huisman

**25** Harvesting carrots: loosen the soil with a fork, then pull the carrots and place in a bin.

**26** Place the harvest in the shade and spray water on to remove field heat.

**27** Washing sunflower shoots.

Credit: Katie Huisman

**28** Dropping the bag of washed greens into the spinning machine.

Credit: Andrew Barton

**29** Sorting sunflower shoots on the drying/sorting table.

Credit: Katie Huisman

**30** Washing loose root veggies; use same technique for loose carrots, radishes, turnips and beets.

Credit: Katie Huisman

**31, 32, 33** A bed can be worked from different positions.

**34** My front yard includes an odd-shaped plot. The sidewalk comes in on an angle and creates an imperfect square.

**35** Short beds: field microgreens in six-foot greenhouse beds.

**36** Double beds: two 30-inch beds side by side. I have forfeited one walkway.

**37** Triple interplanting.

Credit: Katie Huisman

**38** Rototilling a bed.

**39** Using the tilther for no-till bed turnover.

Credit: Andrew Barton

**40** Use a pitchfork to loosen the soil; notice the lack of soil disturbance.

Credit: Andrew Barton

**41** Direct seeding using the Jang seeder; I eye-ball the rows, splitting the difference between each row.

**42** Use a seed-bed roller to mark the holes in which to transplant.

**43** Transplanting tomatoes interplanted in beds of spring greens, forfeiting two rows of greens on the left side of the right bed for one tomato row.

**44** Transplanting kale in landscape fabric beds.

**45** Soil flats in the nursery.

**46** Soil blocks of head lettuce.

**47** Planting microgreens: spread the seed around the flat evenly by hand.

**48** Field microgreens just exposed to light; the board has been removed after the seeds have germinated.

**49** This bed of field sunflower shoots is ready to harvest.

**50** 12′ × 36′ quick poly tunnels built with chain link fencing top rail.

**51** 18′ × 48′ poly tunnel kit.

**52** Poly low tunnels.

**53** Trellised tomatoes using the hard pruning technique.

# Turning a Lawn into a Farm Plot

There are a variety of techniques at your disposal to get a site into production. First, make sure you have gone through and understand the Logistical Land Checklist outlined in Chapter 15 to decide if the place you're looking at will work for your operation. There are a number of factors to take into consideration before you can start developing land into a farm, and these factors will determine the approach you can take.

First, what time of year is it? Are you preparing this plot during a summer to farm the following spring, or are you preparing the land early in the spring to start farming right away? Is this a wet time of the year, or is it dry? During the spring or a wet time of year, if the ground is muddy, tilling can be a mess and can also damage your soil. If it is a hot and dry season, then you will need to take another approach to lessen damage to the soil.

Second, what's growing on that ground right now? Is it a standard lawn growing domestic grass, or is it a plot of invasive grass on a piece of land that hasn't been looked after? These two conditions are also opposites that may govern what approach you'll take. Domestic grass is far less work to deal with, where invasive grass will require quite a bit more elbow grease and time. Don't be discouraged by invasive grasses; most of my plots were covered with it in the beginning.

Third, is the ground really hard packed? If you stick a garden fork into it, how deep will it go? When the ground is really hard, you can't just go in with a rototiller and grind it up. That is actually really dangerous and could cause injury. I strongly advise you not to rototill ground right away if it's hard packed and dry.

For all of these scenarios, I'll explain what the most effective approaches have been in my experience.

### Starting in the Summer or Fall

The most ideal situation is to have a good amount of lead time to prepare your land for the following season. Either the summer or early fall is best in my climate zone. When there isn't as much urgency to get the ground into production fast, there are some passive techniques you can use to make conversion into a farm easier. You can also let time do a lot of the work.

In many North American climates, it's going to be fairly hot and dry during the summer months. If this is the case in your area, using tarps to smother the grass is a very easy and effective way to get the job started. Look for some large, high-quality, black tarps. The ones I have used are thick and black, are about 60 by 40 feet square and cost around $200 each. I have at least five of these on my farm, as I also use them for a pre-emergent weeding technique that I'll explain in Part 7. The first thing you'll want to do before spreading the tarps is to make sure the ground is moist, not dry. Once the ground is nice and wet, cover it completely with the tarps. You may have to water it for a couple of evenings and tarp the areas you've watered so that the water doesn't evaporate during the day. You need to make sure that you have lots of weights along all the edges of the tarps, so that they don't blow away, it's also to help suffocate the grass underneath. I recommend keeping tarps like these on all winter. Once it's warm enough that there is no snow on the ground and not a lot of consistent rain, you can remove the tarps and see if anything has survived. Most likely, everything will be brown. At this point, you don't need to worry about using a sod cutter to remove the grass, and you'll be able to skip ahead and start tilling up the ground.

If you're starting in the summer or early fall and the ground is hard packed and dry, first figure out why the ground is so hard packed. Do this by digging around in a number of places to see what's down there. If the ground is really rocky (there are a lot of big rocks everywhere), you may want to reconsider using the lot. I have turned over some plots with a lot of big rocks, but not absolutely everywhere. It's up to you to make the call if rock removal will be worth the effort.

If the ground is basically dirt but is hard packed because it may have been baked in the sun—and if it has heavy mineral content—then it's possible to use this ground. It just needs to be loosened up. I'm imagining a lot of the red soils I've seen in southern California or Hawaii when I say this. With these dry mineral soils, I'd recommend a time-based strategy, using tarps and a lot of water to loosen up the soil. Heavily water areas of the plot (or all at once, depending on its size) over the course of a few evenings; tarp it over in the early morning to prevent evaporation during the day. This treatment could take up to a week, depending on how hard packed and dry the soil is. After a few days of doing this repeatedly, you should be able to notice a difference in the softness of the soil. Once you've noticed a difference, it might be safe to till the soil. If you can't get it quite soft enough, go in with a broadfork manually or get a subsoiler implement for a tractor in there. A subsoiler is a large hook or hooks that can be pulled behind a tractor; the hooks reach deep into the subsoil and break up the ground. After doing this, hopefully your plot will be safe to rototill.

## Starting in the Spring

If you are starting to prepare your ground in the late winter or early spring and are looking to be in production by late spring, you may need to take a more aggressive approach. In this case, you want to remove the grass as quickly as possible so that you can get right into the soil to start working it. First note that if the ground is too wet or frozen, you won't be able to work the soil until it's somewhat dry and/or thawed. If the ground is really dry too, you may want to wet it down a night or two before you start this process.

Turning a lawn into a farm plot during spring can be done in seven stages. This may have to be done over the course of multiple weeks, depending on whether there is a lot of invasive grass or not.

**Seven Stages to Quick Spring Production**

1. Remove the grass
2. Rototill
3. Form out the beds
4. Rake out debris—rocks, roots and rhizomes
5. Loosen subsoil
6. Add amendments
7. Prepare beds for planting

For the sake of describing how long each task will take, let's assume we're turning over a 2,000-square-foot area.

### *1. Remove the Grass*

Whether this site has domestic or invasive grass, you can start the same way. Go to a local machine rental shop and rent a sod cutter. This might cost around $60 for the day, but it's worth it. Run the machine in straight lines across the lawn, so that you can roll the sod off. For a 2,000-square-foot

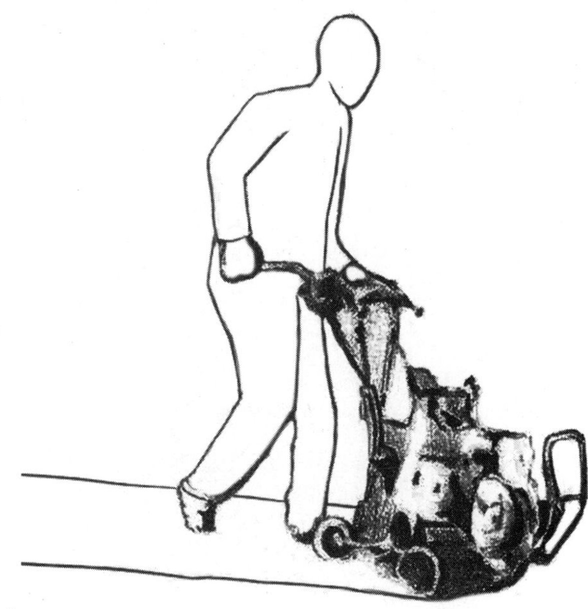

**FIGURE 12.** Cutting sod with a sod cutter.

FIGURE 13. Peeling back the sod layer and rolling it up.

area, running the cutter would take one person a couple of hours. Then, with some help, removing the sod and getting it off site might take another half day. If there is very little or no grass, then you can skip ahead to step #2 and just start rototilling.

FIGURE 14. Walking on the outside of the tilled path forms out the beds as your feet make impressions in the soil.

### 2. Rototill

Once you remove the sod, you can start rototilling the ground. The first pass you make with the tiller will be the most difficult, but each pass will get easier. Make sure you go over the ground multiple times to loosen it all up nicely. It may take an hour or so to till a 2,000-square-foot area for the first time. I usually make at least three passes over the entire area. You want the soil to look really ground up. If you're still seeing large chunks of earth, you'll need to make more passes. For larger plots, of ¼ acre or more, consider bringing a tractor to do the initial tilling. If the ground is uneven at all, you may need to take a landscaping rake to level out some parts that are uneven. Make sure you do this before the next step.

### 3. Form Out the Beds

After the ground has been tilled well, you can start to form out your beds. To make perfectly straight beds, you might have to use string lines as guides. In urban areas, there will often be something to follow, like a fence, house wall or sidewalk. To make beds, you will walk next to your tiller, so you're not stepping on soil you have tilled. With the BCS and Grillo brand walk behind tractors, you can adjust the handlebars to make this easier. The width of the beds you form should be the width of your tiller. If you're not using a tiller that is between 24 and 30 inches, and you'll have to form your beds with a rake. I advise using a 30-inch tiller and going with 30-inch beds; they are the standard size for small-scale, intensive farms. The width you set for walkways is up to you. I set mine anywhere between 6 and 18 inches, based on the size of the site and what I'm going to grow there.

Forming out beds can be done within a few hours once you get the hang of running the tiller in straight lines. Now that the beds are formed, all the work we do will be focused on the beds. Don't worry about anything in the walkways at this point. Those can be managed with a stirrup hoe very quickly.

FIGURE 15. Notice the thumbs up position and how the rake is on an angle, so that debris is raked in a row towards your feet.

#### *4. Rake Out Debris— Rocks, Roots and Rhizomes*

Depending on how much debris is in your soil when you form out your beds, it may be a good idea to let it sit for a few days. After some rain and when the soil settles a little bit, some debris will actually float to the surface. After this happens, it'll be a lot easier to rake things out. If the ground had a lot of invasive grass in it, this process will probably have to be done multiple times. I have also found in the past that, with invasive grass, it sometimes helps to till the beds again, then let them sit for a few days. Then go back and rake the rhizomes out again. If it's really bad, this stage can take weeks. The best way to rake out invasive grass is to move your rake towards you on an angle and walk along the footpath backwards, raking the rhizomes along the edge of the bed into a row. Then, go and pick them up. The key is to keep taking the rhizomes out and to get them off site. Do not put them in your compost pile!

If there are nearby trees and you're discovering a lot of roots, you might have to use a shovel or a pitchfork to pull them out. There have been times when I've had to use a small axe to get bigger roots out of the ground. Small rocks, just like rhizomes, will also float to the surface over time, and this will make it easier to rake them out—or dig them out, in the case of big ones.

#### *5. Loosen Subsoil*

Now that your beds have been worked and cleaned out, you need to loosen the subsoil underneath. Most sites that were lawns

FIGURE 16. Use a broadfork to loosen subsoil.

before will have a certain level of subsoil compaction, especially after you have rototilled. For this job, you can use a strong pitchfork or a broadfork. You're going to need to drive it into the bed as deep as it can go, then pull it back gently. Don't torque it too hard, or you'll break your fork. Do this up one bed, driving it down every foot, and then down the next bed. This can be fairly time-consuming, and you may want to save half a day for this task.

### *6. Add Amendments*

Your beds will look a lot more messy after you've forked them, but that's OK. At this point, I bring in lots of good finished compost and organic fertilizers. There is no particular recipe here, except that my main soil amendment will be compost. The worse the soil is, the more compost I add. I don't use manures, as I find in an urban environment, their smells may bother neighbors. Also, unless the quality is perfect, they can sometimes create more problems then they're worth; weed seeds or some pests like wire worm can survive the digestive tracts of many ruminants. I buy compost from a local business that makes it, and where the soil quality is poor, I'll add a two-inch layer to each bed—and only the beds, not the walkways or perimeters.

FIGURE 17. Using eight-gallon buckets filled with compost, we add two buckets to each 25-foot bed.

### *7. Prepare Beds for Planting*

At this point you could either rototill again or use a tool such as the tilther to mix your amendments in. One pass over each bed with a tiller or tilther is all you would need at this point.

## Starting Any Time with Large Plots

If you are preparing an area that is ¼ acre or larger, you may want to consider bringing in heavier machinery to speed the process up. Rototilling a large plot of raw land with a walk behind tractor can be really backbreaking and take more time than is necessary. If you have lots of open access on the land, consider hiring someone with a tractor for an afternoon to do the first three stages. With ¼ acre or more, the stages could go as follows:

1. Remove the grass with a front end loader or tractor
2. Loosen subsoil with a tractor
3. Till with a tractor
4. Form out the beds with walk behind tractor
5. Rake out debris—rocks, roots and rhizomes
6. Add amendments
7. Till the beds again to prepare for planting

When you're preparing a plot this size, most of these stages are the same as for small plots. For stage #1—if you have enough access, you could use a front end loader or tractor to remove the grass. Use the bucket to scrape off the top one inch or so. You could also get a tractor to loosen the subsoil with a subsoiler implement. If the plot is in a peri-urban area and there are other farmers around, see if you can pay someone to come in to do the initial tilling. A full-size tractor can do it much faster than a walk behind tractor.

# Choosing a Site

## Your First Site

The first farm plot that you develop is important, but don't get too hung up on finding the most perfect place. The main idea is to find a place where you can get some production going, so you can gain experience and income and build social equity in your neighborhood or city. The beauty is that once you start, more land will find you, because people will see what you're doing, they'll ask questions, and then you've got them engaged. That's what you want.

The first step is always the hardest, but once you're in motion it gets easier as you go. Your first site might be either a small piece of land to farm—or it could be your base of operations—or ideally, it could be both. Over the years I have cycled through 20 pieces of land and have moved my home base three times. That sounds difficult, I know, but over time it has led to more permanent and long-term leases on land that I may have never found if I hadn't started fast and built a name for myself in my city.

If you're looking for a home base (a place where you can store tools, keep infrastructure, and work from), you want to have it in an area where there are other pieces of land nearby. To start, your home base might be your neighbor's or uncle's garage, with no land on site to farm. You might store your walk-in cooler, tools and other processing infrastructure there. Once you have spotted a potential place like this, it would be ideal to find another piece of land nearby that you could farm, and the potential of other sizeable lots of land close to this. Going onto some mapping software might be the best place to start. Look for an area of town that you'd like to be in, look at the amenities nearby, the farmers market, potential restaurants and most importantly, look for homes with big lawns. If you can find one place that has good visibility in a neighborhood like that, it's guaranteed you're going to have people in the neighborhood approach you once they see you out there working.

A perfect scenario would be to find a piece of land that had a place that you could farm (maybe 2,000 square feet or more) as well as a place to set up your home base infrastructure. The important thing, however, is to get some production going; that will lead to many other opportunities.

## Satellite Farms

Once you have your first site found, you're going to want to expand to other plots. The key here might be to wait a little bit: once you get some production going on your first site, let other landowners come to you. To expedite this process, I recommend putting up a sign on your first plot, even when it's still under development, saying what it is and how people can get in touch.

Over the years, my satellite plots have moved closer and closer to my inner circle. I attribute this to the fact that I spread the word about what I was doing; by now I am connected to many people in my city. In my first couple of years, I had plots scattered all over the city; some were five miles away or more. But over time, I kept getting offers for land closer and closer to where I

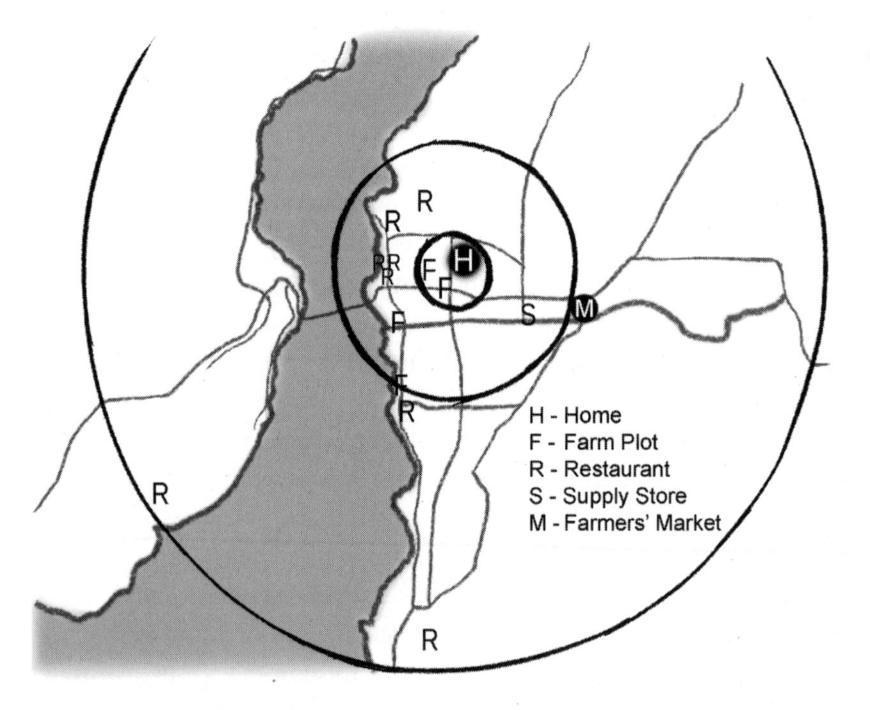


**FIGURE 18.** This map illustrates how my farm, customers, farmers market and suppliers are laid out within the city. The three rings represent the Thunen rings from Figure 1. Credit: Anthony Ross.

wanted to be. I did leave many plots behind, and in most of those cases, the landowners became avid gardeners afterwards and were happy to take the land back. On some occasions I seeded grass back for the landowners, but that happened on only a couple of occasions.

Where each satellite plot is located in relation to your home base will determine what kinds of crops grow there. For the plots furthest away in the network, you'll want to focus on Bi-Rotation (BR), but not just any BR: only crops that don't need a lot of maintenance. Tomatoes and pattypans, for example, are considered BR, but tomatoes need constant pruning, and pattypans need to be picked at least every second day during summer to get them at a small size which makes them more valuable. High maintenance BR crops still need to be in your inner area. Crops like kale, carrots, beets, scallions, chard, peppers and eggplant are the BR crops that can be farther away in your network. Once crops like these are established, they can be harvested once a week. By using landscape fabric and compostable plastic mulches, once the plots are set up you should go there only to harvest. Very little other maintenance is required.

# Garden Layout

## Standardization

Whether a plot is in HR or BR, there are some things you should consider regarding how the plot is laid out. Plots in HR are primarily growing Quick Crops, and since their turnover is fast, Quick Crops often don't need the same amount of space to grow in as some of the Steady Crops in BR areas. For example, walkway width between beds for crops like radishes, leaf lettuce and other leafy greens can be narrow—sometimes as tight as six inches or even less depending on how nimble you are on your feet! At full maturity, Quick Crops don't grow so high in foliage that they drop out over the walkway. On my farm, areas in HR will have six- to eight-inch walkways. Plots in BR will in most cases have at least one-foot-wide walkways. When plants like kale are at full maturity, for example, the foliage will spread out into the walkways a little bit, so you need to make sure that they're wide enough to accommodate that.

## Factors Affecting Layout

With most small farms, the layout of the beds will often be based on the track the sun follows, drainage or trade winds. These are important things to consider, but for urban farms, other factors come into play that can greatly affect production on a site. The major factors to consider when establishing a farm plot are shade obstructions (such as buildings, fences, trees), but most importantly, the shape of the plot itself will determine the best way to lay out your beds.

As often as I can, I try to run my beds at 25- to 50-foot lengths, but sometimes that's not possible. My main consideration is to maximize growing space and minimize the areas that are less important such as walkways and perimeters. The main reason to have a standard length for beds is consistency when it comes to irrigation, season extension equipment and any pest protection row covers. The problem with having beds of different sizes and shapes is that odd sizes can cause a lot of headaches when you're looking for the right size piece of row cover. Unfortunately, compromised layouts are just a reality if you're farming on multiple urban lots. So, the best you can do is to have as much consistent bed length as you can.

## Sunlight and Shade

Understanding the track of the sun in each time of the season and each time of the day is critical when deciding how to configure the layout of your farm plot. Refer to the land checklist in Chapter 15 (Scouting for Land) for details.

# The Perimeter

The perimeter of a plot is very important. When you're establishing a new site, you need to consider where your access points are, and how you can best set the site up so that you're not wasting valuable time maintaining areas that are not productive.

## Access Points

Figure out early on how you get in and out of the plot; you'll want to make sure you've thought of all possible scenarios before you start to lay out beds. The main things to take into consideration are fences, gates, existing walkways, hedges, trees and where the house sits in relation to the plot. Make sure there is enough width in the access point(s) you choose to move in and out whatever machinery (e.g., a rototiller) you're going to be using. Just imagine that you're going to be coming in and out of this plot day in and day out for years at a time. Before you lay out any of your beds, first determine what the path of least resistance to get in and out would be.

## Sheet Mulching

On sites where I have wide pathways as access, I will sometimes lay down wood chips to keep weeds down, and to have a nice area to walk on during the wet months. On some plots where I have long-term tenure, I will sheet mulch the entire perimeter around the beds, especially areas where constant work is being done. On my home base plot, all the areas that are walked on all the time are sheet mulched with wood chips. I have found that it's much better to lay down chips than it is to have grass. Though grass may look nice for a while, it will eventually get mucky after it has been trodden down, especially where water is being used for washing.

Spreading wood chips can be hard work though, and it's pretty time-consuming. If you don't have long-term tenure on a plot of land, going through all that work is probably not worth the effort. Wood chips can often be found and delivered for free if you call some local arborists and ask them where they dump their chips. I find that the fall is the best time to get free chips delivered to you. This is usually when a lot of tree pruning happens in the city, and these arborists would rather drop chips off somewhere in town than drive all the way to a distant city refuse site.

### Landscape Fabric

Landscape fabric has become my preferred choice to cover areas of exposed soil for the sake of smothering weeds. What I like about it is that it's fast; it's easy for one person to cover an entire perimeter in an hour or less. I can even take the fabric up and move it to a new location if I'm only at a plot for a couple of years. If you buy the high-end material, you can use it for up to three years or more if you treat it right. I used to spend a couple of hours each week during the summer using a string trimmer to knock down large weeds and grass around the perimeters of my plots until I realized what an unnecessary waste of time this was. First, it was causing problems: the debris from the trimmer would end on my greens, thus making me have to wash them. Second, I was wasting around 20 or more hours a season doing this. Now, I lay down three- to six-foot-wide landscape fabric around the perimeters of every one of my plots. If there is a walkway access point between segments of beds, I will put down fabric there as well. This means that I no longer waste time running a line weeder or forking out weeds in areas that have gotten out of hand. This is a proactive approach, and will greatly reduce your labor inputs. See a landscape fabric perimeter, photo #6, in the photo insert section.

# Irrigation

There are two main forms of irrigation I use on the farm: overhead and drip. For the most part overhead is used in Hi-Rotation areas, and drip is used in Bi-Rotation areas. In my greenhouses I sometimes use both types: overhead micro sprinklers for greens and drip for tomatoes or summer crops.

I use overhead in the HR areas because these beds are turned over so often that I don't want to have to move drip lines between three and six times a season, every time I replant. I am in a high desert, which is a very dry climate for Canada, but it's not nearly as dry as some other places in North America; we do get lots of precipitation during the winter. In hot dry climates (like some southern US states), using drip would be a must, even for HR areas.

## Decide on Your Approach

The type of irrigation you should use has a lot to do with the crops you're growing—whether they are crops that occupy a small footprint on the ground, or they're crops that cover the entire bed. In both of these cases, drip or overhead can be used, depending on whether the plots are HR or BR areas. Tomatoes and summer squash, for example, are planted in straight rows and occupy only a single line in a bed; there's only need to water close to the crop. Otherwise, you're watering bare ground and encouraging weeds on unproductive areas. For these crops, you could use a single line of drip irrigation, applied just to the base of the crop. But for a crop that occupies more of the bed, you will need to use more lines. On a 30-inch-wide bed, I have found that three to four lines will cover the bed evenly. When I'm planting Steady Crops such as kale, beets and sometimes carrots, I will use this method. I have found that if you are direct seeding crops of anything beyond four rows, four drip lines will give better coverage. If crops such as beets, kale or chard are being transplanted to three to four rows, three lines of drip will work fine.

For greenhouse irrigation, I will sometimes use a bit of both drip and overhead. In all of my tunnels, I use micro sprinklers that follow the top ridgepole and are spaced three feet apart. When the tunnels are planted with Quick Crops in the early or late part of the season, this configuration of sprinklers waters everything in the tunnel. Then, as I transplant tomatoes, I will run one drip line across the base of those plants. I will continue to run the overhead sprinkler for the first couple of weeks (when the greens that are planted are being transitioned out), but when the greens are gone, all the irrigation will be running to the drip lines on the tomatoes alone.

### Know When to Water

When you water your crops is very important, and it will change during the seasons.

#### *Early Spring/Late Winter (Cold Season)*

I water my first plantings in the ground only during the early spring when the crops need it: mainly when they are planted, and possibly only once a week after that. When temperatures are cool, there is far less evaporation, so less need to water. Also, during the month of March in my area, we still get freezing temperatures at night so I don't want to run my irrigation systems and risk the lines freezing and bursting. For these times, I hand water the first crops that are planted in the greenhouses or select segments on farm plots.

#### *Spring (Cool Season)*

For the main part of the spring, I run my irrigation in the mornings or after greens have been harvested. Since there isn't a high level of evaporation during this period, I still don't need to water every day. If you don't water before you harvest greens, you can avoid washing them, which is a huge time-saver. I like to water in the morning during the spring because I want a little bit of evaporation so that the soil stays warm; this way I can avoid overly wet soil that might encourage slugs or fungus. At this point in the season and all the way until the fall, my irrigation systems are all automated. These timers can be changed weekly and be adjusted for different cycles for each day of the week.

#### *Summer (Hot and Dry Season)*

During the hot months, I run water every day during the evening. Usually as the sun is setting we will experience a little bit of wind as the temperature drops. I don't water then, as wind can blow the water around and cause uneven coverage. I usually wait until dark. By watering overnight you get the most amount of time for the water to soak into the ground without any evaporation. Sometimes, on very hot days where temperatures will rise to 100°F, I will run sprinklers for ten minutes at the hottest point of the day, especially on areas that have just been seeded, to encourage germination. I will often hand water specific beds that were just seeded during this time as well, and sometimes cover the beds after being planted with some kind of row cover until they germinate. In hot and dry climates, you need to be careful when planting during your hottest months. Fast evaporation will cause spotty germination that reduces yields. I have about eight weeks of very hot weather in my climate, and I do use considerably more water during that period.

To decide if soil needs water, I stick my finger into the soil; if it's dry for the top two inches, I'll water. After I run a watering cycle, I'll go back later in the day, or the morning after watering, stick my finger into the ground again, and if it's wet only on the top inch, I know I need to water longer. Testing often, in this way, is particularly important during the summer months.

#### *Fall (Cooler Season with Shorter Days)*

As my growing season moves into fall, I will start to scale back my watering. I might go back to three days a week, or even less; it really depends on the weather. If temperatures are below 50°F during the day, in most cases, I won't water unless the soil looks like it needs it. As overwinter crops like lettuce and spinach get seeded, I make sure to water these thoroughly to encourage strong germination. Once they're germinated, there's no need to keep watering if the temperature is cool.

#### *Winter*

In my area, crops rarely need water in the winter. The only time I would water is if I were establishing a very late winter crop in the greenhouses for early spring. In this case, I'd water at the beginning of the day, but only if it's above freezing, and just enough to keep the soil moist to encourage good germination. Crops like these would be spinach, lettuce, asian greens, radishes or salad turnips.

### Drip Systems

Drip is the best approach when farming on plots in Bi-Rotation. Most crops that grow at these plots are in the Steady Crop category, and are usually planted as plugs. (Two exceptions to this would be carrots and beets during the warmer months when they are direct seeded.) Drip irrigation only waters the crop. Kale for example is planted at ten-inch centers with three rows in a bed; this way, I will need only three lines of drip. If you are noticing that your drip lines aren't spreading over your beds evenly, you may need to add another line.

I use a technique called a *flow through system,* and this means that there are no dead ends anywhere on the lines. The end of each bed has a main line that feeds all the way through. Running the lines this way means that if one of the lines is ever clogged, you will never have dead sections because water can flow in both directions. For example, if you run lines as single rows, from one direction, and there is a clog somewhere on that line, everything after the clog won't get watered. However, if water is coming from both directions, everything has a better chance of even watering. See photo #12 (irrigation lines) in the photo insert section.

- Some drip is called *pressure compensating,* and this type is better suited for running drip along slopes or really long runs, because it drips the same amounts to each emitter, no matter where that emitter is located (on top of a hill or down at the bottom). If your plot isn't on a slope, then you needn't worry about using pressure-compensating drip lines.
- I prefer drip that has emitters every six inches opposed to twelve inches. Using three or four six-inch lines on a 30-inch bed will give far better coverage. Twelve-inch spacing can work, but you will need to stagger the emitters on the bed, aligning emitters on the two outside lines and setting emitters on the middle line off center. This will create a more distributed coverage. In really sandy soil, which drains quickly, you may need to go to four lines per bed.
- I prefer to use hand-tightened couplers for all my drip lines as well. I like to be able to unfasten the lines on a bed with just my hands. Hand-tightened couplers do cost a little bit more, but you make the savings up by the ease with which you can undo the lines to turn over a group of beds.

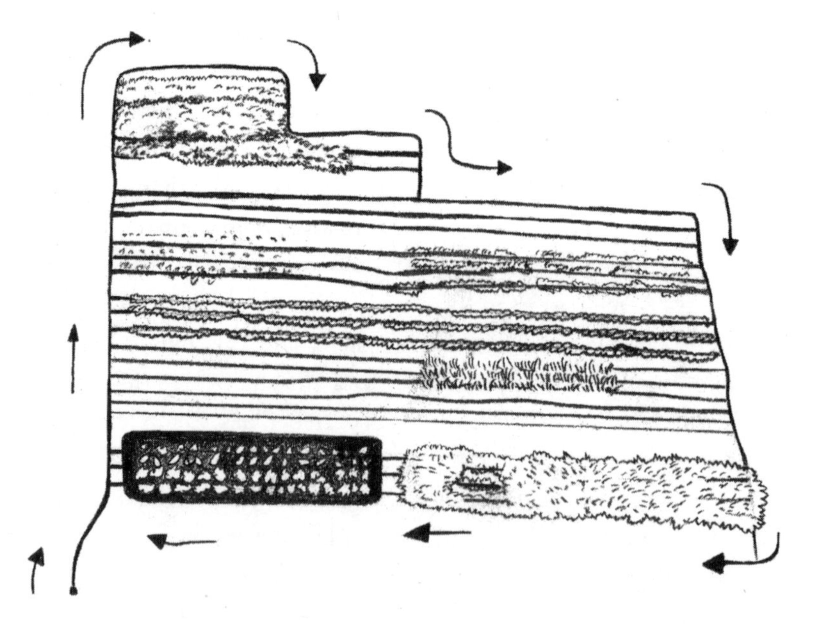

**FIGURE 19.** The arrow shows flow through a drip irrigation system.

**FIGURE 20.** Drip Irrigation Budget for 1,000-Square-Foot Plot (Ten 25-Foot Beds).

| Drip Equipment | Used For | Item Cost | Needed for one 10' × 25' bed (approx. 1,000 sq. ft. area) | Total Cost |
|---|---|---|---|---|
| ½" poly tubing, 100' roll | main lines that connect to the drip lines and to the timer, pressure reducer and filter | $15 each | 60' | $15.00 |
| Tee-Smart-Loc™ coupling | connects the main line to the drip lines | $2.50 each | 60–80 | $150–$200 |
| 90° elbow Smart-Loc™ coupling | on the corners of the plot | $2.50 each | one for every plot corner (3) | $7.50 |
| Straight Smart-Loc™ coupling | connects main lines or severed drip lines | $1 each | only needed in some cases | — |
| ½" drip line, 6" emitters, 1,000' | drip lines on the bed; 3-4 lines per 30" bed | $150 each | 750'-1,000' | $150.00 |
| Filter | filters particulates, makes drip lines last longer | $40 each | one per plot | $40.00 |
| Pressure regulator | reduces the pressure from your source, most drip lines need to be reduced to 12 psi | $15 each | one per plot | $15.00 |
| ½" male threaded to ½" male poly coupling | connects pressure regulator to the filter and main line | $2.00 | 4 | $8.00 |
| ½" male poly to ¾" female threaded garden hose coupling | connects the timer to the beginning of the line | $2.00 | 1 | $2.00 |
| ½" hose clamps | fastens the poly tubing | $0.90 | 6 | $5.40 |
| 1-4 zone timer | garden hose timer | $50–$100 | 1 | $50–$100 |
| **Total cost for 1,000 square foot area** | | | | **$442.90–$542.90** |

### *Overhead Systems*

I water all of my plots in HR with an overhead system using standard impact heads. The key to establishing even watering here is to make sure you have head-to-head coverage. This means that water coverage of each impact head will not overlap but reach the one next to it. I make all of my own stands out of cut two-by-fours and poly tubing. It's simple, reasonably cheap and very effective.

On a plot that is a perfect square (say 50 by 50 feet), you need to make sure that each head is dropping the same amount of water as the next, even if they are running at slightly different angles.

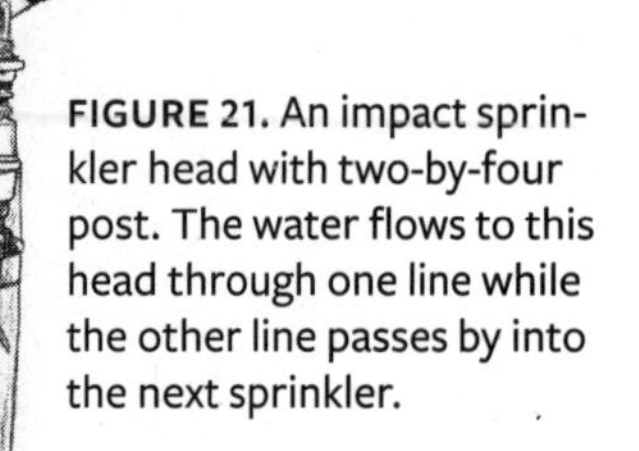

**FIGURE 21.** An impact sprinkler head with two-by-four post. The water flows to this head through one line while the other line passes by into the next sprinkler.

It's best to have the impact heads operating in each zone using the same direction of spray. For example, on a zone that has two heads spraying water at a 90° angle, if that zone were to run for 15 minutes, another zone that had two heads running at 180° would run for double the amount of time, because it is covering twice the area. I follow this is the basic rule of thumb. See photos #7, 8 and 9 in the photo insert section.

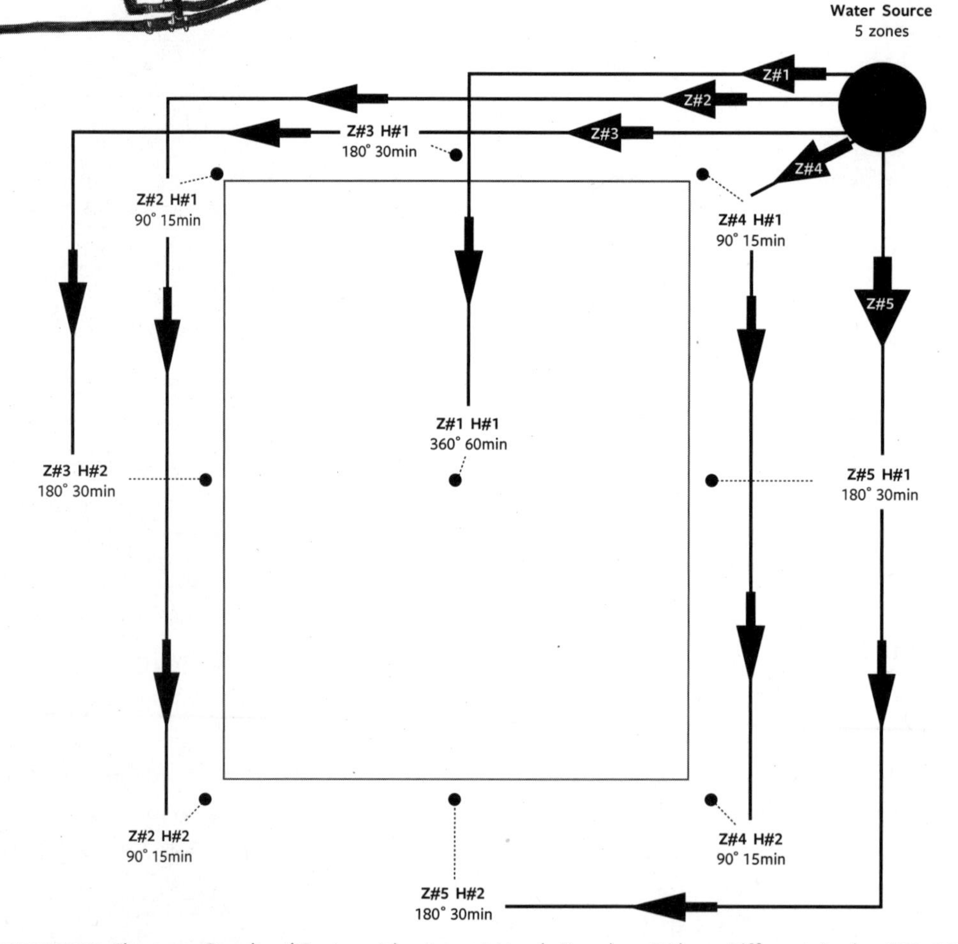


**FIGURE 22.** Five-zone Overhead System: Nine Impact Heads Running at Three Different Angles: 90°, 180° and 360°. Credit: Curtis Stone.

**FIGURE 23.** Irrigation Materials List for Twelve 50-Foot Beds

| Overhead Equipment | Used for | Item Cost | Needed for one 12 × 50' bed (approx. 2,500 sq. ft. area) | Total Cost |
|---|---|---|---|---|
| ¾" poly tubing, 100' roll | main lines that connect to the sprinklers | $28.00 | 400' | $112.00 |
| Tee, coupling | connects the main line to the drip lines | $2.50 each | 4 | $10.00 |
| 90° elbow coupling | on the corners of the plot | $2.50 each | 5 | $12.50 |
| Straight ¾" coupling | connects main lines or severed drip lines | $2.50 each | only needed in some cases | — |
| Filter | filters particulates; makes impact heads last longer; put the filter on before the timer | $40 each | one per plot | $40.00 |
| ¾" male poly to ¾" female threaded garden hose coupling | connects the timer zones to the lines. | $2.00 | 5 | $2.00 |
| ¾" hose clamps | fastens the poly tubing | $0.90 | 50 | $45.00 |
| 4 zone timer | garden hose timer | $100.00 | 1 | $100.00 |
| 1 zone timer | garden hose timer | $50.00 | 1 | $50.00 |
| 2 × 4 × 12 for sprinkler posts | cut in half, used for 6' posts for sprinklers: treat the bottom 2' with a waterproof stain, and bury 2' into the ground with sand or rocks | $5.00 | 5 | $25.00 |
| ¾" female threaded to ¾" poly, for posts | for sprinkler posts: ¾" male poly to ½" threaded from poly line to ½" poly coupler | $2.50 each | 4 | $10.00 |
| ½" poly female to female threaded coupler | for sprinkler posts: connects the impact head to the male threaded from the poly line | $2.50 each | 9 | $22.50 |
| ½" threaded poly impact head | the sprinkler | $10 each | 9 | $90.00 |
| Irrigation box or cover | to keep sun off the timers | $60.00 | 1 | $60.00 |
| **Total cost for 1,000 square foot area** | | | | **$579.00** |

On a site that is not a perfect square (see photo #10 in the photo insert section), it can get a little bit more complicated, but the same concepts for head-to-head even coverage still apply. For the first zone of sprinklers here, we will run one line from our main water source to zone #1-head #1, through to zone #1-head#2, and end at zone#1-head#3. All of these are on the same zone, because they are all spraying at 90°, and they can be run for the same time (15 minutes). Zone 2 will run zone#2-head#1 and zone#2-head#2 together. They are both spraying at a 180° angle, so keeping them on the same zone is important to get even coverage. In this case, we'll run zone 2 for 30 minutes, double the time of zone 1, because zone 2 heads are covering twice the area. Zone 4 is odd because it is going to cover the center area, and it'll be spraying

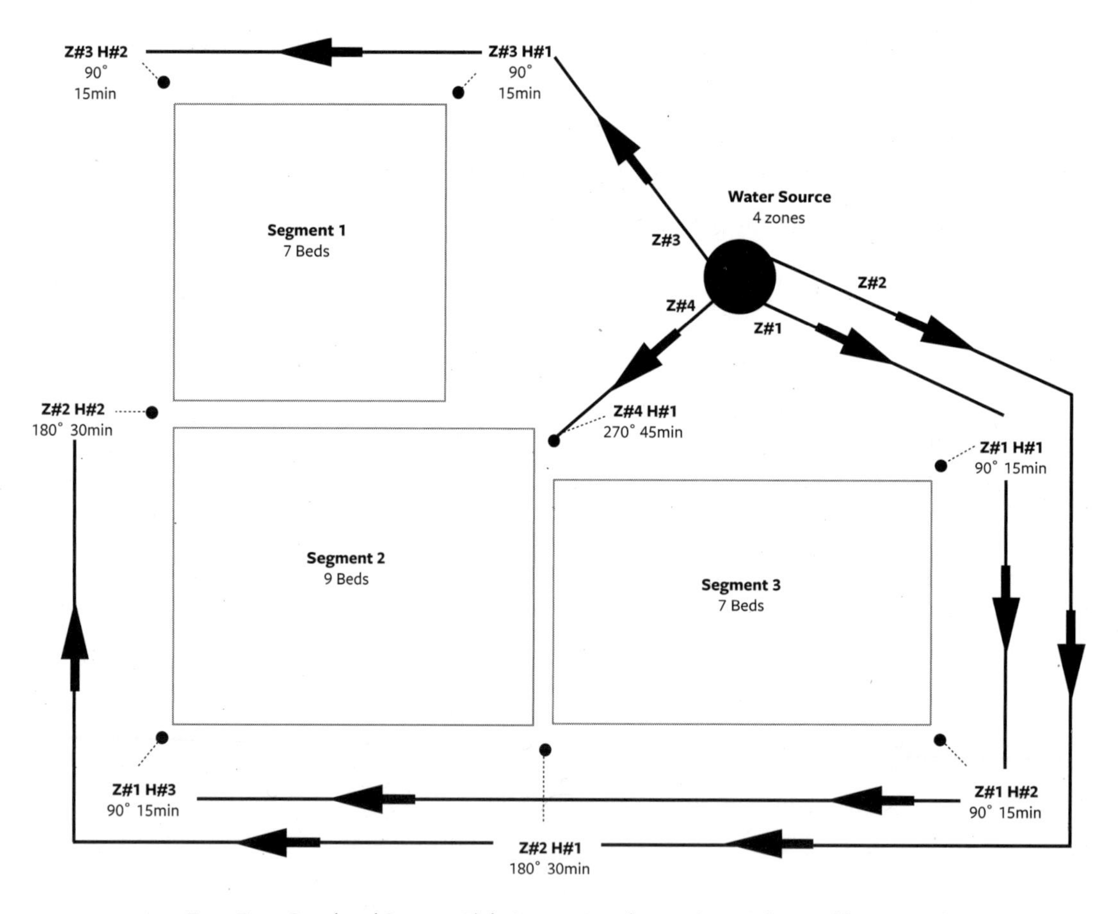


FIGURE 24. Four-Zone Overhead System: Eight Impact Heads Running at Three Different Angles: 90°, 180° and 270°. Credit: Curtis Stone.

at 270° angle. Since zone 4 is one of a kind on this plot, it will have only one head. This head should be able to reach all the corners of the plot. Since the zone 4 head will be covering three times as much area as zone 1 at 15 minutes, it will run for 45 minutes. In zone 3, we have two more heads at 90°, so they will both be run for 15 minutes (just as in zone 1).

When building your system, it's very important to set how much water each head will drop so that you are not kicking up soil onto your crops as water hits the ground. If the density of the spray is just right, it won't splash water onto salad greens. When you are shopping for the right impact head to use, find something that has three adjustments: spray density, arc and angle. It's best to have all three. Once the system is all tied together, I will turn on a zone, and spend a few minutes at each head, tweaking the angle, spray and arc to be just perfect. You want to make sure that each head is reaching the next one, and that the spray isn't kicking up dirt from the beds or pathways.

### *Timers*

I have installed irrigation timers at all of my farm plots. This is critical, because to

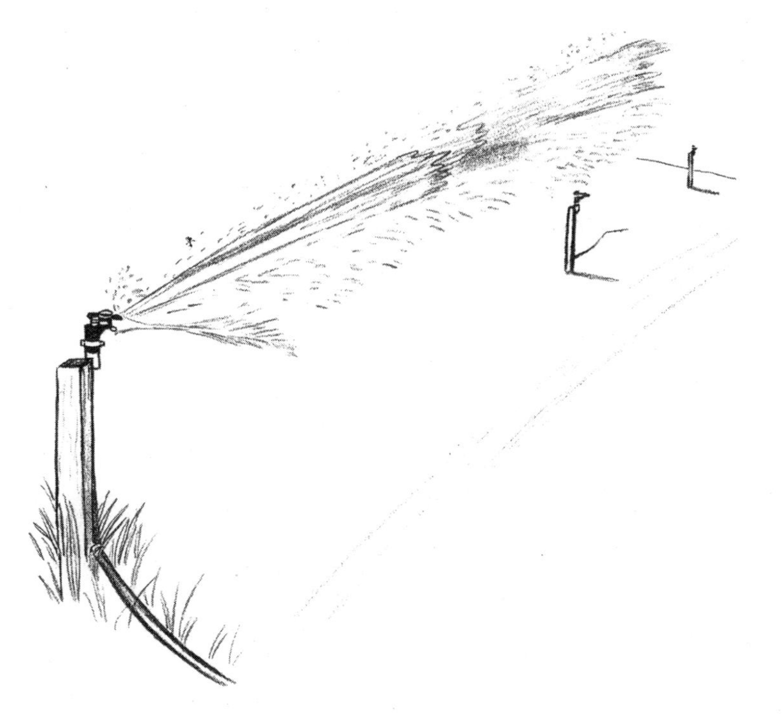

**FIGURE 25.** Each sprinkler must be able to reach the next one in the line, and overlap a bit. This is called head to head coverage.

travel from plot to plot and manually turn on and off water would be a huge waste of time. I use simple one- to four-zone timers on all of my plots. These are an inexpensive option and can be found at most hardware stores. They are the only piece of irrigation equipment that I use that doesn't always come from an irrigation specialist outlet. See photo #9 (an irrigation box) in the photo insert section.

### *Greenhouse Systems*

Greenhouses can be irrigated in two ways, and the best approach really comes down to personal preference. In my large tunnels, I have a micro sprinkler system that includes one main line running along the top ridgepole, with sprinklers set every three feet. These sprinklers spin at 360°, at head-to-head coverage. See photo #11 (micro sprinklers) in the photo insert section.

**FIGURE 26.** Irrigation equipment list for a generic small greenhouse (18' × 48'): one main line is zap strapped to the center ridgepole, with micro sprinklers tapped in every three feet. The watering of the sprinklers is meant to spread head to head, and will reach the edge of the tunnel.

| Greenhouse Equipment | Used for | Item Cost | Needed for one 18' × 50' greenhouse | Total Cost |
|---|---|---|---|---|
| ¾" poly tubing, 100' roll | main lines that connect to sprinklers | $28.00 | 1 | $28.00 |
| Filter | filters particulates, makes impact heads last longer; put filter on before timer | $40 each | 1 per plot | $40.00 |
| Pressure regulator | reduces pressure from your source; most drip lines need to run at 12 psi | $15 each | 1 per plot | $15.00 |
| ¾" male poly to ¾" female threaded garden hose coupling | connects the timer zones to the lines. | $2.00 | 1 | $2.00 |
| 1 zone timer | garden hose timer | $50.00 | 1 | $50.00 |
| ¾" hose clamps | fastens the poly tubing | $0.90 | 10 | $36.00 |
| Micro sprinkler with anti-drain valve | valve keeps sprinklers from continually dripping when the system is turned off | $1.50 | 16 | $24.00 |
| Hanging sprinkler | tubing is punctured and hangs from the main line; 1 sprinkler every 3', hanging down a few feet; can be cut or shortened to desired height | $1.50 | 16 | $24.00 |
| **Total cost for 1,000-square-foot area** | | | | **$219.00** |

PART 6

# INFRASTRUCTURE AND EQUIPMENT

Many of the tools and equipment that an urban farmer needs have to be compact, mobile and lightweight. Not only are we dealing with a lack of space to store things, but we're also farming in small spaces with tight corners, so even tools have to be easy to operate under those circumstances. Infrastructure also needs to be non-permanent, because it may need to be moved from time to time.

# Base of Operations

It is best to have your base of operations in one place. It's ideal to have it close to or at your home, or close to your main network of plots. Your base of operations will be the place where you have a workstation for processing and packing produce, your walk-in coolers, a place to store your tools and machinery and even an office (though that could be in your home). The home base for my farm moved three times over the first five years I was operating. Each time, I moved it closer and closer, until it ended up being at my home (a place that I was farming one year, then came to rent the next, then purchased two years later). Now my base of operations is in my backyard and carport. Having it at home is very convenient because I also have my office on the same property. Don't be discouraged if you can't find the most ideal situation at first. More opportunities will appear after you've been operating for a while, once you've built a strong network of customers and supporters.

I like to lay out the main parts of my work area according to how often I use them and how they fit into the general flow of a workday. My walk-in coolers are right in my carport and easily accessible because they are the most used piece of equipment. When product comes in from the field, it goes from my truck or bike trailers directly into coolers; then it's pulled out for processing and portioning, then put back in afterward. Then product is unloaded for deliveries to restaurants or farmers markets. So to have the coolers in an easily accessible place is very important.

I keep my most used tools near the entrance to my work area. I will often load up my truck or bike trailers with tools that I'll need for the day, near the entrance to my carport, so I like to have my tools close by.

The components of a work station that are used in conjunction are the washing station, spinning machine, drying table and portioning area. They must all be close to one another because in your daily workflow you'll be using all of them together. It's ideal to have these stations relatively close to your walk-in coolers as well.

Where your office is in this mix isn't as important. However, if you are using an invoicing system with software, you need to have that close to your work area. If your office is at home, and your work station is somewhere else, printing your invoices could be an issue. I prefer to print invoices for our restaurant customers after the orders have been put together, just in case we're short on some things and we have to modify the orders a little bit. In the past, before my office was at my home base, I kept a printer at the work station and used my lap top for invoicing.

**FIGURE 27.** Major purchases for an urban farm: some of these may not be necessary (depending on the type of farm you choose to operate), and you may already have some of this equipment (transportation and office supplies).

| Base of Operations and Major Purchases | Approximate Investment Cost |
|---|---|
| Cooler (buying used) | $1,000.00 |
| Washing station | $100.00 |
| Spinning machine | $70.00 |
| Drying table | $60.00 |
| Portioning station | $630.00 |
| Office | $0–$1740 |
| Tools | $1,550–$7,050 |
| Farmers market equipment | $980.00 |
| Nursery equipment | $788.00 |
| Indoor vertical nursery (optional) | $865.00 |
| Greenhouse nursery | $500.00 |
| Poly low tunnels 6 tunnels (12 beds) | $415.80 |
| Quick tunnel (optional) | $500.00 |
| Transportation (may already have) | $0–$3,000 |
| **Total investment range** | **$7,458–$17,698** |

## Coolers

A walk-in cooler is a vital piece of equipment you must have on a farm. There are three reasons you need a walk-in cooler; they pertain to quality, marketing and workflow:

1. When product is harvested, you must remove the *field heat*, to stop it from spoiling. Some crops (particularly greens or microgreens) are more susceptible to heat degradation than others. To maintain a high standard of quality, you must remove the field heat.
2. Marketing. You gain a considerable amount more time to market any product when it is stored properly. You don't have to sell it the next day. You can buy days to weeks of time to sell your harvest, depending on the type of product.
3. By using coolers, we can harvest on multiple days on the farm. For Friday restaurant deliveries and a Saturday market, we start harvesting on Wednesday. Instead of harvesting everything the day before, you can spread it over a few days. This helps keeping a more streamlined workflow, by accomplishing harvesting at the same time as other tasks.

Because my farm is small compared to most other operations, and because I've had to move my base of operations a few times, I prefer to have two medium-sized walk-in coolers rather than one large one. Lack of

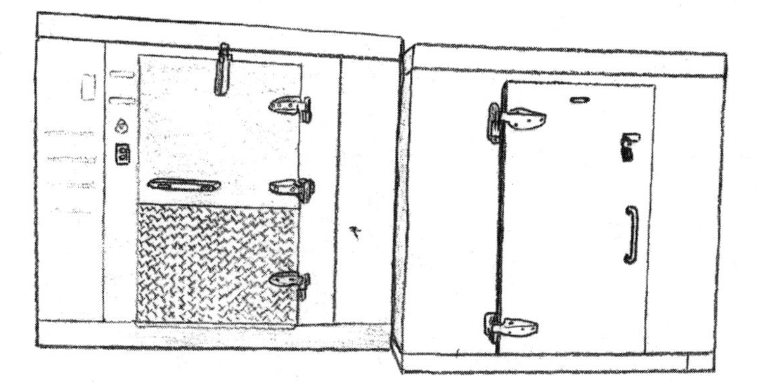

**FIGURE 28.** Our two, medium-sized walk-in coolers.

space is often an issue for urban farmers. The nice thing about having two coolers is that only one of them is running constantly. The other runs only on Thursdays and Fridays when we are packing for restaurants and the farmers market. We use one for produce that was just harvested, and the other is for produce that has been washed, portioned and packed. The two cooler system keeps things organized and easy to find.

I've seen some farmers use a collection of home refrigerators as coolers. I would not advise doing this if you're planning to farm commercially. It's very difficult to fit containers from the field into domestic refrigerators. The space inside is too small. This means you have to process and portion everything immediately when it comes off the field so that it will fit in the coolers, and this is not efficient workflow. Some farmers will use commercial pop coolers (you see these in gas stations or grocery stores). These can work for small operations, but you may need a few of them. Sometimes, you can find ones that have enough inside volume to fit harvest containers or bins. That would be the key. If the coolers can't hold your totes, baskets or bins, then don't buy them. The main problem with these is that if the compressor malfunctions or dies, they're not worth fixing. Heating and cooling repairs are very pricey.

I recommend restaurant-style walk-in coolers or custom-built coolers using a CoolBot™ system with an air conditioner for a compressor. These cooler choices are both economical and easily available. You can find restaurant-style walk-ins online or at auctions, or simply find restaurants that are liquidating their assets. The restaurant industry has one of the highest failure rates of all businesses, so the turnover of equipment is quite high.

The next best option is to build a cooler yourself. There are many designs online. You can frame in a simple room, any size you want, but the larger you go, the more cooling capacity you need. A standard size

**FIGURE 29.** Recommended cooler specs for four farm sizes: look for a cooler with just the panels, whether or not the unit has a working compressor. By using a CoolBot™ system, you can modify a $200 air conditioner (far cheaper utility costs).

| Land Size | Minimum Cool Storage Area Needed | Cubic Feet | Cost Range |
|---|---|---|---|
| ¼ acre | 4' × 6' × 6' | 144 | $1,000–$2,000 |
| ⅓ acre | 4' × 6' × 6' × 2 | 288 | $2,000–$4,000 |
| ½ acre | 6' × 8' × 8' | 384 | $2,500–$5,000 |
| ½ acre and beyond | 8' × 8' × 8' | 512 | $3,000–$6,000 |

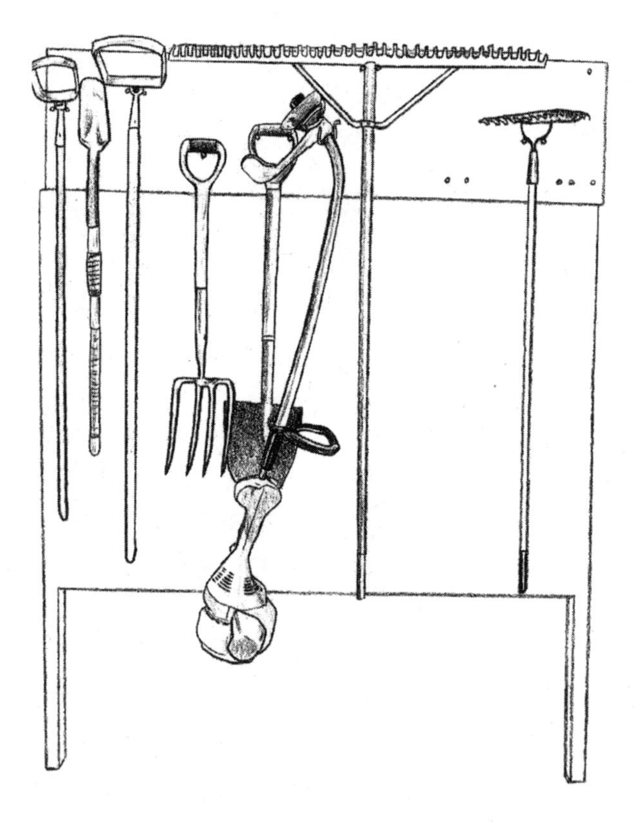

**FIGURE 30.** It's a good idea to hang tools up off the ground, so you are not tripping on them, especially in high-traffic areas.

for a farm, if you have room for it, is eight feet wide, eight feet deep and six feet high. There are lots of options for high-grade insulation and vapor barriers to choose from. The key to this design is the cooling unit itself. Instead of buying a compressor, you can use a simple piece of electronics called a CoolBot™; it modifies an air conditioner to act as the compressor.

It can be pandemonium on a farm if your cooler fails in the middle of the summer. One season, the compressor on my restaurant cooler died, and to fix it was far too expensive. So I ordered a CoolBot™ and rigged up an AC to the top of my cooler, using some styrofoam to direct the cold air into the cooler. I was back and running in half a day. These units also use less power than full-size compressors.

### A Place for Tools

Keeping your tools organized and in their proper place is very important. Having to look for a tool every time you need one adds up to a huge waste of time over the course of a season. I prefer to keep my tools hanging up and off the ground. Then they are out of the way, so I'm not tripping on them or getting smacked in the face by a rake. I keep all of my hand tools near my workstation, hanging from simple hooks. I also have a small toolshed on the back of my property where I store my rototiller, chipper/shredder, flame weeder and any other machines or less commonly used tools.

## Washing Station

The washing station is where I wash products such as radishes, carrots, scallions, turnips and beets. The washing station is built out of two-by-fours, ¼-inch steel mesh and pool liner. The table itself is a very simple design, and the mesh is screwed on using wide washers. There are no sharp edges anywhere on it, so that the table doesn't rip or tear things. The pool liner wraps around the bottom; this directs water into a large plastic tote, and then it is pumped away, directed to some berry bushes and trees on my property. See photo #30 (washing loose root veggies) in the photo insert section.

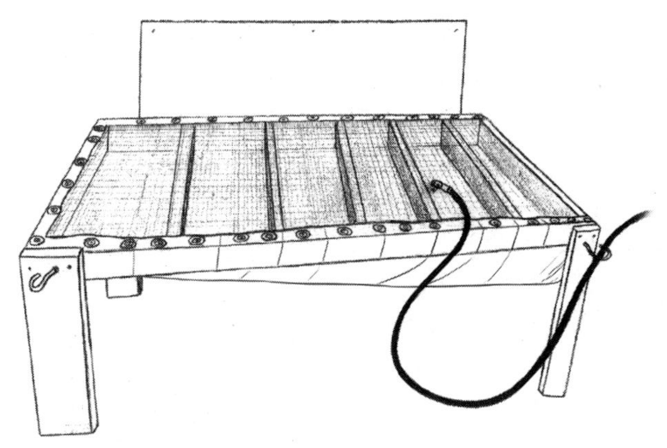

**FIGURE 31.** Washing station: 34″ wide × 95″ long × 31″ high. Materials: 8′ × 3′ ¼-inch steel mesh; five 2 × 4 x 8s, two 2 × 6 × 8s; 30 mil PVC pool liner (60″ × 100″); 100 × 1.5″ wood screws; 50 × 1″ wide washers for wood screws to hold down pool liner. Approximate cost $100.

## Spinning Machine

The best way I have found to spin greens is to use a modified washing machine. I realize that in some jurisdictions this is not allowed, but this is what has worked for me. If you are producing anything near 100 pounds of greens per week, you're going to be wasting labor turning a hand-powered spinner. In my first two years, I was using a hand spinner to spin dry our greens, and I was cranking one of these things for four hours a week, not an efficient use of my time.

I use two pieces of equipment to spin greens: a modified washing machine and a few laundry bags to hold the greens while they are in the machine. I purchased this machine for $50 from an appliance repair shop, and I asked them to make the cone in the center shorter so that my laundry bags, when filled with wet greens, fit better into the machine.

See photo #28 in the photo insert section.

When they're being washed, the greens are placed into the bags to drip out some of the water. Once a bag is full, it goes into the spinning machine for five minutes. From there, the bag comes out and the greens are dumped onto the drying table. The purpose of using the bags is twofold:

1. It is easier to move delicate greens from station to station, reducing the amount of handling
2. Using bags reduces the amount of debris left in the washing machine, making the machine easier to clean

## Drying Table

Once the greens come out of the spinner, they go onto the drying screen. This is a simple table built with two-by threes and ¼-inch mesh. There are two box fans that hang from the top of it. The base of the table below the mesh drops out when you are finished drying one set of greens; this makes it easy to clean off leftover debris. It's connected by two hinges to the back of

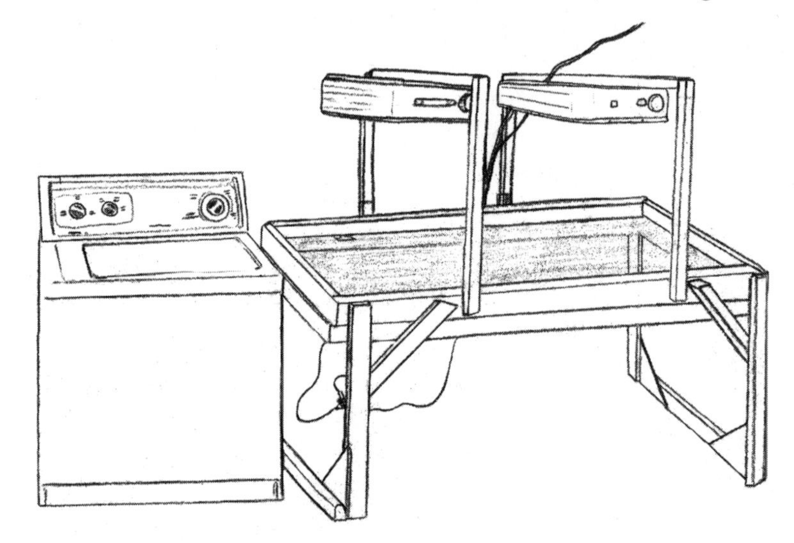

**FIGURE 32.** The modified washing machine costs approximately $50, laundry bags $10 each.

**FIGURE 33.** Cost of materials for a three-by-six table = $60.

**FIGURE 34.** Equipment Recommended for Portioning Station

| Portioning Station | Item Description | Approximate Investment Cost |
|---|---|---|
| Analog scale | one analog scale for weighing bins | $100.00 |
| Digital scales | two digital scales for portioning | $60.00 |
| Collection of bags | (covered in detail in Part 7) | $200.00 |
| Twist ties | for closing bags | $20.00 |
| Elastic bands | some items may be bunched while portioning, like kale | $20.00 |
| Shelving unit | a place to store bags and scales | $50.00 |
| Tables for packing | possible to use market tables | $80.00 |
| Bins for packing | bins, totes or crates for packing | $200.00 |
| **Total** | | **$730.00** |

the table. See photo #29 in the photo insert section.

I find that greens, and especially microgreens, have a longer shelf life once they have been dried. The process only blows off any residual water on the greens; do not dry them to the point of wilting. Once the greens have been spun, they may sit on the table for only five to ten minutes, depending on the type of green.

## Portioning Station

The portioning station is the place where all the bagging and bunching takes place. This area has a shelving unit where I keep scales, bags, twist ties and elastic bands that we use to assemble orders and pack items for the farmers market. This area is on the deck of my house. It's covered to provide shade and shelter from the elements, and it's fairly flexible in how it's laid out. Depending on how many people will be helping with portioning on a Friday, I can set up multiple work areas to accommodate more people by setting out fold-up tables I use at the market. There could be anywhere from one to three people helping on a Friday, so I have enough space here to allow for that. This area is close to my coolers, as helpers will be taking finished product from a cooler, bagging or bunching it, putting it into a labeled tote and then back into a cooler. Since I have two walk-in coolers, I use one for unfinished product and one for finished product. This makes things easy to find.

## Office

Having an office is important, but not totally critical at the beginning. If you are going to focus more on serving restaurants, then having an office space will be more important. The main reason is that you will be invoicing the restaurants. Very few will be paying cash. You'll be producing a lot of paper invoices that are signed and collected. I print two copies of an invoice, one for the customer to keep and one which they sign and I take away. When I come back after deliveries, these signed invoices go into a wall inbox named "delivered invoices;" later I file them by the month.

The main components of my office are

- desktop computer $1,000
- accounting software $500
- laser printer for invoicing $200
- inboxes on the wall $40

The cost to set up an office can vary greatly. Since most people already have access to a computer and printer, and may already have a home office, this expense may not be necessary for most.

I use my accounting software for invoicing, tracking sales and open (unpaid) invoices. When you're dealing with a lot of restaurant customers, this is very important. It can be easy to lose track if you don't have a good system. With my software, I can list all of open invoicing (the customers who owe me money). Most customers have 21- to 40-day terms, meaning that they must pay after a set amount of days. Once I'm into peak season (the month of May and after), I have checks coming in every week. Every Saturday, after the farmers market, I put all of these checks into the system. If I have noticed that some customers are behind in payments, I make a statement of how much they owe, and I e-mail it to them. The statement is a list of all the unpaid invoices. It's important to have a system for doing this so that people don't get behind in paying.

Only on a few occasions have I experienced customers not paying invoices. For this, my protocol is as follows

- If someone is two weeks behind their terms, I send them a friendly and polite e-mail stating that they are behind in payments. If they are a new customer, we state that we can't deliver any more product until the previous invoices have been paid. If they are an older customer, we are generally more forgiving, especially if it's not something that happens often.
- If some customers are consistently behind in paying and I have to badger them every time to get paid, I stop delivering to them. This rarely happens.

# Tools

For the urban farmer, hand tools are the primary tools we use. Since urban farming is more or less where square foot gardening meets commercial farming, a lot of tools we use will be common to both gardeners and farmers. Save the tractor. Even the machinery on my farm could be used in a private garden.

All farms will have some of the classic tools such as pitchforks, shovels, picks and rakes. All these are used for harvesting, digging or shaping ground.

### *Pitchfork*

The pitchfork is used to harvest carrots, weed by digging out the roots of deeply grown weeds and to loosen soil. Sometimes I prefer using a pitchfork for loosening subsoil. In very rocky ground, it can be hard to drive a broadfork down deep enough. When I want to loosen subsoil where the side walls of my greenhouse are too low, I will use a pitchfork.

### *Broadfork*

The primary function of this tool on my farm is to loosen subsoil for no-till bed preparation. It can also be used for harvesting carrots, but the walkways on my farm

FIGURE 35. Caption: If you're on a tight budget, it's possible to start without buying all of these tools in your first year. QCGH, lawn edger, broadfork and walk behind tractor are not essential to start.

| Tools | Approximate Investment Cost |
|---|---|
| Pitchfork | $40.00 |
| Broadfork (not essential) | $200.00 |
| Stirrup hoe | $60.00 |
| Lawn edger (not essential) | $50.00 |
| Landscape rake | $50.00 |
| Walk behind tractor (possible to rent or buy used) | $0–$5000 |
| Planting equipment | $600.00 |
| Harvest bins | $500.00 |
| Harvest knives | $50.00 |
| Quick Cut Greens Harvester (not essential to start) | $500.00 |
| **Total** | **$1,550–$7,050** |

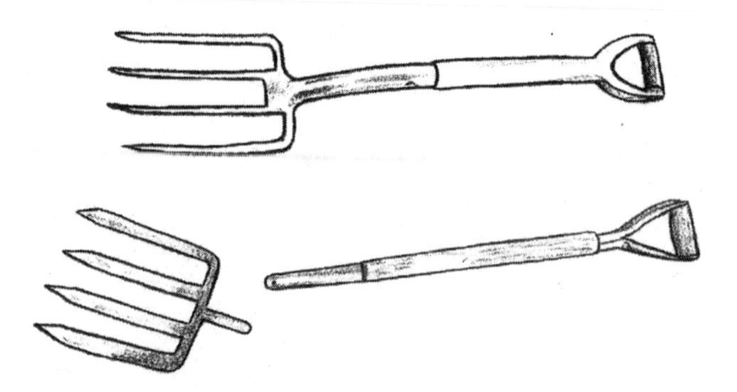

**FIGURE 35.** If you want a pitchfork that lasts, buy one where the fork is connected to the shaft with one solid piece or weld. Pitchforks that have the fork connected at the end and are held in place with screws, bolts or glue will not even last a season.

are often too narrow to use it effectively. Broadforks can also loosen the ground to pull out invasive grass or weeds when preparing a new site for production (see figure 16 in Part 5).

#### *Stirrup Hoe*

The only real weeding tool I use on the farm is a stirrup hoe, and it's mostly used for weeding walkways and perimeters. I personally prefer hand weeding within garden beds. I also use a stirrup hoe as a no-till bed-turning tool.

#### *Lawn Edger*

An edger is a common landscape tool that is used to cut edges of grass around a perimeter. I use it for exactly that, and it serves no other purposes. It's not a critical tool to have, as you can use a shovel in a similar way.

#### *Landscape Rake (Bed-Preparation Rake)*

The landscape rake is a pretty important tool on the farm and is essentially a bed-preparation rake. It's three feet wide, and I use it for a couple of different bed-preparation tasks: spreading compost evenly and shaping beds when I'm turning them over without a tiller. I also use it a lot during plot preparation, especially if the ground has a lot of invasive grass in it. This is the tool we use to rake grass rhizomes out from the beds (see Part 5, Figure 15).

### Rototiller and Walk Behind Tractor

For most of the years I have been farming, my walk behind tractor has been critical to my farm. However, in recent years as I have moved to more no-till methods (as outlined in Part 8), I believe it is possible to run a farm of ⅓ acre or less without owning one. In this case, you could rent one when it is needed for plot preparation. If you are farming on more than ⅓ acre, owning a machine is a worthwhile investment. There are many implements to choose from, but the most important one to start with is the rototiller. If you can find one of these units used, it is worth it; if these machines have been looked after well, they will last a lifetime. A new a machine with a tiller can go for $5,000, but used you may be able to find one for $1,000. See photo #38 (me using a tiller) in the photo insert section.

### Planting Equipment

Proper seeding equipment is a must on a commercial farm. Don't even try to plant seeds in the ground by hand. Some farmers think that since they are operating at nearly a garden level, they can get away with hand seeding. Please save yourself the wasted hours and buy a seeder!

For an entry-level seeder, you can start with an Earthway for $100. It's simple and has some flaws, but it works. The most precise seeder I have seen on the market is called the Jang Seeder. It has a lot more options for precision seeding at varying densities, but will cost you around $600. It is now the main seeder we use on the farm. Unlike the Earthway, it uses rollers instead of plates to pick up seed. The rollers I use for the Jang are all described in Part 10. See photo #41 (direct seeding) in the photo insert section.

### Harvesting Equipment

For harvesting baby greens, I used a simple handheld knife for years. I started by using a small, serrated steak knife for all lettuce, arugula, mustards and spinach. I encourage you to learn good knife and hand techniques before using high-tech tools such as The Quick Cut Greens Harvester (QCGH), because there will always be times when you need to harvest by hand. Carrying QCGH from plot to plot all the time is a lot to lug around, but keeping a small knife in your truck or bike pouch is easy. The QCGH has changed the game on our farm: what used to take eight hours can now be done in 45 minutes. That's a massive time savings; even though this unit costs around $500, it paid for itself within the first month in reduced labor cost. If you are not producing over 50 pounds of greens per week, this tool may not be necessary. See photos #21 and #22 in the photo insert section. Harvesting equipment is covered more extensively in Chapter 32 (Harvesting).

### Farmers Market Equipment

When you're preparing to attend a farmers market, like most things in farming, it's important to buy the right equipment. If your market is outdoors, the most common way to set up is to use some kind of market canopy. A ten-by-ten-foot square is the most common size that I've seen for market vendors, but there are a lot of different sizes out there. Do not buy a cheap market canopy. Spend more to get the best one you can. This is something that you will set up and tear down hundreds of times over, and if you get something cheap, you'll be replacing it in less than a season.

FIGURE 37. Recommended Market Stall Equipment: spend a little extra to get the best quality; you want things that will not break while you're at the market.

| Farmers Market Equipment | Use | Approximate Cost |
|---|---|---|
| Fold-up tables | Look for good quality | $50.00 |
| Market canopy 10' × 10' | This creates your stall at the market. Bring weights to hold down your canopy: milk jugs filled with water, tied with bungee cords work well. | $500.00 |
| Tablecloths | | $50.00 |
| Signs | Get a banner that is easy to see and large enough that people can see it from a distance. | $150.00 |
| Display boxes | These can be purchased new. Sometimes old fruit packing boxes can have a rustic look. | $150.00 |
| Item tag stands and clips | Hold your labels and price tags | $50.00 |
| Cash box | For making change | $30.00 |
| **Total** | | **$980.00** |

Fold-up tables are easy to find at most department stores, and you will need a couple of these for your display. The nice thing about these tables is that you can use them for other tasks on the farm, like packing and sorting for market prep. These tables also will see a lot of use, so buy good ones.

Other items you'll use at the market stall are tablecloths, labels for produce items and display boxes or bins. Retail supply stores will often have all you need for labels, as well as other ways of displaying things at your market stand. You can also make these accessories yourself if you're crafty. Some if the best-looking stalls at markets are handmade; these stalls have an authentic and down-to-earth look which people at markets love. See photo #1 (my farmers market stall) in the photo insert section.

# Special Growing Areas

## Nursery

The nursery is the place where all early season planting takes place. For many urban farmers, it can be a logistical challenge to find a place to put a greenhouse, especially in a multi-location context. However, I have created some simple structures that can serve as small, moveable nurseries. When I started, the first nursery I built was a small 12-by-20-foot greenhouse with vertical shelves to maximize the space. See photos #13, 15, 16, 45 and 46 in the photo insert section.

Besides the basic structure to set up these nurseries, the other equipment needed are things such as 10-by-20-inch flats of various sizes. I use 200 and 128 sq. cell flats, 2½-inch and four-inch pots, germinations trays, soil mix, compost, soil sifter, a soil mixing table and an all-purpose organic fertilizer.

## Indoor Vertical Nursery

This setup can be built using steel wire shelving (commonly used in restaurants) or industrial shelving with fluorescent lighting installed. I use shelves that are two feet deep and four or eight feet wide. You can change the length to however many trays you can fit in. The most common light fixtures are T8 fluorescents with two bulbs each; each fixture is four feet long. So, on one two-by-four-foot shelf, I would have two light fixtures with four light bulbs total. I have found that you need four lights in order to get adequate light to the crop.

You can keep this type of setup in a heated garage or even your kitchen. If you have large south-facing windows, this type of nursery can also be set up there. I do find that natural light grows a better crop in general, but it depends on how much light reaches in your house, and how sunny

**FIGURE 38.** Recommended Nursery Equipment: the soil sifter, germination flats, watering wand and potting soil that are used in the nursery will also be used for microgreens.

| Nursery Equipment | Use | Amount | Cost |
|---|---|---|---|
| Mini soil blocker | Used to start lettuce, beets, spring onions and tomatoes from seed. I fit 420 blocks in a 10 × 20 germination flat. | 1 | $30.00 |
| Germination flats (1" × 10" × 20") | Mainly used for microgreens, but also for mini soil blocks | 1 box of 100 | $200.00 |
| 128 SQ flats | 128 cells per flat. I use these mostly for early kale, chard and sometimes beets. | 1 box of 100 | $200.00 |
| 2.5"–4" pots | We pot up tomatoes from small plugs into 2.5" pots and sometimes 4" pot if they can't get out into the soil right away. | 1 box of 500–800 | $70.00 |
| Soil sifter | Use this to make a fine soil mix. Build with a simple frame from 2 × 4s and ¼-inch steel mesh that straddles a tote. Push soil and compost through the screen to make a fine mix for potting. | 1 | $30.00 |
| Soil mixing table | Build with 2 × 4s and plywood. It's a waist-high table for making soil mix. | 1 | $80.00 |
| Potting soil | There are many brands out there that will do the same thing. I use an organic peat-based mix with perlite. Buy 4 bales to start. | 4 | $100.00 |
| All-purpose fertilizer 4/4/4 | I use an all-purpose NPK rating of 4/4/4 for all of my general nursery potting. | 1 | $28.00 |
| Soil mix for soil blocks and pots | To make soil mix: 1 part screened compost to 2 parts potting soil. 2 quarts fertilizer to 210 quarts of soil mix. I mix 3 bins of 70 quarts each at a time. | — | — |
| Watering wand | Use a wand with a soft and low pressure spray. | 1 | $30.00 |
| Brass on/off valve | Connects to the wand. Buy one that is all brass, with a lever that is at least 1.5" long. | 1 | $20.00 |
| **Total** | | | **$788.00** |

it is in your area throughout the winter, or whenever you do your first nursery planting out.

Watering your flats can be a little challenging if the vertical nursery is in your house. In the past, I filled up a small plastic tote with a few inches of water and then placed the flats to be watered, one at a time, right into the tote. I let the flat sit for a few seconds, then lift it out, let the water drain out then put the flat back on the shelf. Keep in mind that this type of nursery is not the most ideal, but it works and is easy to build for a small investment. For an urban farmer on ½ acre or less, mainly focusing on high value, this can work for many years because such a small farm does not require huge amounts of transplants. On my farm, I need only about 32 ten-by-twenty-inch flats to get the season's crop started. Once things warm up a little outside, I will then move things out to the greenhouse.

**FIGURE 39.** Materials List for Indoor Nursery

| Indoor Nursery Equipment | Use | Amount | Cost |
|---|---|---|---|
| Chrome wire shelving (2' × 4' × 6') | Vertical shelving units for micro flats. Four 1020 flats per shelf level. A 6'-high unit could have 4 levels 18" apart, holding 16 flats. | 1 | $150.00 |
| Soil tamper | Cut a ¾" piece of plywood to 9.75" × 19.75". Use this tool to press down soil in flats to make a firm planting surface. Screw a small door handle on the back so it can be picked up easily. | 1 | $10.00 |
| Soil sifter | To make a fine soil mix: build with a simple frame from 2 × 4s and ¼-inch steel mesh that straddles a tote. Push soil and compost through the screen to make a fine mix for potting. | 1 | $30.00 |
| Germination flats (1" × 10" × 20") | Mainly used for microgreens, but also for mini soil blocks. | 1 box of 100 | $200.00 |
| T8 Fluorescent light fixture | These are the main lights for this nursery. For a four-level unit, use two per level for eight total. | 8 | $160.00 |
| Light timer | A simple timer to turn the lights on and off. This is connected before the power bar, where all the lights are plugged in. | 1 | $20.00 |
| Dehumidifier | It is critical to keep humidity low when growing indoors to prevent mold. | 1 | $200.00 |
| Stand-up fan | Constant airflow is very important when growing indoors. | 1 | $60.00 |
| Power bar | Heavy duty power bar with 8 inputs | | $35.00 |
| **Total** | | | **$865.00** |

In figure 39 I list all the main materials to install this kind of nursery. This nursery can also double as a small microgreens production system (more on microgreens in Part 8). See photo #14 (indoor vertical nursery) in the photo insert section.

You can pack a lot of flats into a small backyard greenhouse if you make good use of the space by optimizing vertical areas. I've done this in two ways:

1. With a small 200-square-foot nursery, I built shelves out of wood, and even put fold-up tables on top of other tables to create more vertical space. I built this nursery for $500 in materials. A small greenhouse nursery can be built with the same materials as the quick tunnels or hoop houses discussed in the next part.
2. I've also used a poly tunnel covering ground crops as a nursery. I hang steel Ts from the ridgepole and lay two-by-fours on the T-bars to hang flats from the top.

During colder months, the sun is lower in the sky, and in both of these cases very little shade was cast on the plants beneath.

Sometimes, I have also isolated a small portion of a poly tunnel to make a small nursery. This is basically a greenhouse within a greenhouse.

I don't use any of these options until the outdoor temperature is above freezing at night. I try to keep all of my early starts inside my indoor nursery for as long as I can until I run out of space. That usually happens when tomatoes get moved into larger pots. See photo #15 (poly tunnel nursery) in the photo insert section.

# Inexpensive Season Extension

For urban farmers, season extension in the field can be a logistical challenge. Sometimes the shape and location of a plot can prevent you from installing greenhouses, not to mention the fact that we often don't own the land or have long-term tenure. Building greenhouses can also be very expensive, and this may be a cost a new farmer can't afford right at the beginning.

I use three types of poly tunnels on my farm: poly low tunnels, quick tunnels and hoop houses. Low tunnels are the cheapest and fastest to set up and use, and the other two are slightly more expensive, but offer advantages and are easy to use.

## Modular Poly Low Tunnels and Row Cover

Poly low tunnels were the first type of field-applied season extension I used, and I still use them today. The tunnels are made with greenhouse plastic that cover two

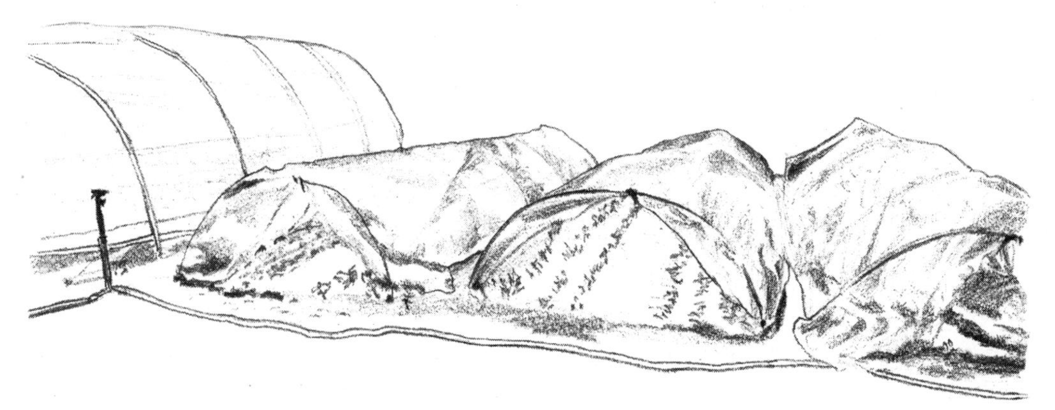

**FIGURE 40.** One-inch spring clamps hold open the ends of these three tunnels.

30-inch beds. The structures are made with ten-foot lengths of ½-inch electrical metallic tubing (EMT), bent in the shape of a half circle and driven into the ground one foot deep. The tunnels can be as long or short as you like. I typically use them for 25- or 50-foot beds. The metal hoops for the tunnels are tied together using a nylon rope that is anchored through a rebar stake with a hoop, both at the beginning and at the end of the tunnel. I tie a knot on the rebar stake, and then pull the rope tightly to the first hoop, and then another knot there. Once these are taut, the rope wraps around each hoop until the end, then another two knots are tied between the stake and the last hoop. The rope acts as a ridgepole to prevent the tunnels from collapsing from snow or heavy wind. Spring clamps and rock bags are used to hold down the plastic. The spring clamps are also used to hold the ends of the plastic open for ventilation during the days.

To make these tunnels, I first purchase a 20-by-100-foot roll of greenhouse poly, then find an open area where I can roll out 35-foot lengths at a time. I use a 35-foot length for a 25-foot bed. You need to have an extra ten feet because the tunnels are a few feet off the ground, and you need the extra length so that the plastic can reach the ground. Roll out your poly, and cut it to your desired length; then, with very sharp scissors, split the poly down the middle. A 20-foot-wide roll of poly will be folded into quarters, and splitting it down the center is simply a matter of following the seams on the folds. My base cost for these units is $63

**FIGURE 41.** Materials Needed to Make Poly Low Tunnels

| Poly Low Tunnels | Use | Amount | Cost |
|---|---|---|---|
| Low tunnel bender | Can be purchased or easily built out of wood. Use to bend the EMT hoops. | 1 | $60.00 |
| ½" x 10' EMT | The hoops for the tunnels. Use 6 per 25' bed; $2.30 per hoop. | 6 | $13.80 |
| Rebar metal stake with loop | Two stakes for each tunnel hold the rope down which acts as a ridgepole. | 2 | $6.00 |
| ¼" twisted poly rope | Rope is anchored to the rebar loops, tied to the starting hoop, then looped through each hoop and tied again to the end hoop and anchored. Use 40 feet per 25' bed. 1,200' roll costs $45; cost = $1.50 per bed. | 1 | $1.50 |
| 20-foot-wide 6mm greenhouse poly film | A 20' roll can be split down the center to make two 10' wide pieces for PLTs. 20' × 110' roll costs around $215 and makes 6 tunnels; cost = $35 per tunnel. | 1 | $35.00 |
| 1" spring clamps | To hold open the ends during the day, and to secure the poly in windy areas. 6 clamps cost $4; I use at least 6 per 25' bed. | 6 | $4.00 |
| Sandbags | Sandbags can be filled with sand or rocks to hold down the poly as well. Cost = $1.50 each; 6 per 25' bed. | 6 | $9.00 |
| | Cost per tunnel is $69.30. That's $35.65 per 25' bed. | | |
| **Total** | | | **$129.30** |

for the poly, hoops, clamps and ridgepoles. So that's around $31.50 per 25-foot bed that uses 6 EMT hoops per bed. See photo #52 in the photo insert section.

### Quick Tunnels

A quick poly tunnel is a simple and cost-effective season extender that can be built in a day or two. The base materials are a ten-foot chain link fencing top rail, ground posts, 6mm greenhouse poly, channel lock and two-by-fours. These tunnels can cover up to four 30-inch-wide beds with eight-inch walkways, and can be built to whatever length you prefer. On my farm, I've never had a plot longer than 50 feet. One 50-foot quick tunnel costs around $500. The advantage of these greenhouses for urban plots is that they are relatively small and can fit into most urban backyards. See photos #42 and #50 (quick tunnels) in the photo insert section.

### Hoop Houses

The hoop house is a more traditional type of greenhouse. In the backyard of my house, I have two 18-by-48-foot hoop houses. They were purchased used for less than $1,000 each. A hoop house is more like a traditional greenhouse. When I bought mine, all the pieces were prefabricated with the ridgepoles and hoops. I custom built the end walls and installed a channel lock to hold down the poly. The only time I would suggest building a hoop house is if you have long-term tenure or secure lease or you own the land yourself, because of cost and the time it takes to set the hoop house up. See hoop house photos #11, 20, 37 and 51 in the photo insert section.

# Transportation

Transportation may be one of the biggest up-front investments you make on your farm, or perhaps not one at all. However, because urban farms are small, huge investment is generally not required. Since we're offering a small amount of product with variety, we don't need the same size vehicles that most farms do. For farms operating on ½ acre or less, a small truck, minivan, station wagon, SUV or bicycles with trailers are all you need to make your day-to-day delivery to restaurants and farmers markets. My initial investment in transportation was a little over $1,000 for a bicycle and custom-built trailer. I didn't have an electric assist on the bike when I started, which presented some challenges.

## Trucks and Vans

Small trucks, stations wagons and minivans can work fine for an urban farmer. I run a small three-cylinder import truck, and it meets all of our daily needs on the farm. I can take everything I need to market, move around my rototiller and pick up 1.5 yard loads of compost to take to my urban plots. Many people getting into urban farming may already have a vehicle that they can use. A large truck or van is not totally necessary. See truck photo #17 in the photo insert section.

## Pedal Power

When I started my farm all of my deliveries to market and restaurants were completely pedal powered. This had some challenges on the physical side, as it meant that I was often spending time and energy riding back and forth from plots when I could have been doing more constructive things on the farm.

The advantages of farming this way for me were that it was a relatively cheap way to start up, but the best part was the amount of publicity it created. People still today know us as "the bike farmers," and we have a very dedicated following of people who firmly believe in what we are doing. When you're riding around on a bike and trailer, pulling a 400-pound rototiller, you're sure to turn some heads. Every single day I sell at the market, someone comes up to me who said they saw me biking around with the trailer, and they are always really excited about that.

In order to run an urban farm by bikes, you need to choose the crops and locations very carefully. You need to minimize your travel times between plots as best as you can, because it can eat up a lot of time in a day. You can bike ten minutes to one plot and then realize you forgot a tool or do not have enough space on your trailers to bring everything back that you need. This will easily eat up a lot of time in your day, and can be very counterproductive. My best advice for running your farm with pedal power is to farm only plots that are within a ½ mile of each other, and to grow only crops that are lightweight. If I had to start over again, that's what I would do. Once your farm grows and you want to expand, only take on more distant plots of land if you can purchase a vehicle. The Hi-Rotation and Bi-Rotation growing systems can also assist you; the old saying "Necessity is the mother of invention" is absolutely true in this case.

We now use two electric-assist, longtail utility bikes with custom-built steel-frame trailers. These bikes cost around $3,000 each with the electric assist, and the trailers were $400 to get a local welder to build. Not that cheap, but maintenance is far less than for a motor vehicle when you factor in insurance and wear and tear. See bike photo #18 in the photo insert section.

PART 7

# OPERATIONS

On any farm, many different tasks need to be performed on a day-to-day or weekly basis. Your goal as an urban farmer should always be to minimize, streamline or eliminate all tasks that don't have any immediate, quantifiable return, so that you can focus on tasks that deliver a measurable return: harvesting, planting and marketing. The least amount of time possible should be spent on tasks such as weeding, crop thinning, turning compost and hand watering. Minimize these tasks so that you have more time to spend with your loved ones, enjoy life and ultimately be more productive. I work on average 48 hours a week on my farm, and in the summer, even less.

# Work Smarter not Harder

You have probably heard the old cliché "Work smarter, not harder." In my spare time during the winter, I play a lot of the racket-based game called squash. It's a great example of this principle. During my younger years when I was learning this game, I'd play with some older men and women who played very strategically and would have me running circles around them. I would get so exhausted that I would lose my ability to think straight and control the ball, and thus, I'd lose the game. Acting strategically is what working smart is all about, and in the farming context, this means using time, technique and the appropriate technology to better leverage your labor.

Being aware of your time is very important. When I started the farm, I timed every single task that I did to help give me better understanding of how long things should take. I documented them all. By knowing how long tasks take, you can better plan your week. On Sunday evenings after dinner, I will usually spend 30 minutes planning my week. I have a list of all the various tasks that need to be done, and I put them into a calendar. Since I know how long certain tasks will take, I can get a pretty clear image of what the week will look like. For any task that takes me a certain amount of time to do, I will usually give 15–20% more if someone else is going to do it.

List all of the tasks on your farm. Note the ones you do well and enjoy doing. The tasks that you don't enjoy or don't do that well should be the first ones you delegate to others, at least sharing them between yourself and someone else. On my farm, I don't particularly enjoy bagging greens; not only do I not enjoy it, but I know that my time can be more effective in other places. So that was first thing that I started to use help for.

One task that I really enjoy and do very well is planting, so to this day I still do most all of the planting on my farm. I will often get help with early transplanting, but when it comes to succession planting and direct seeding, I do all of that myself. It's a task where there is very little margin for error. Eventually, I will need help with it, but it may be one of the last tasks I delegate. I'm not suggesting you don't delegate a task like this, but you should definitely prioritize what can be best delegated first.

### Main Farm Tasks

- Washing
- Portioning
- Packing
- Irrigating
- Harvesting
- Planting
- Office work
- Recording Field Data

Another way to make good use of your time is to evaluate everything you do based on return and cost. I used to spend a couple of hours a week turning compost piles. This work is menial labor (as far as I was concerned). One, I didn't enjoy it at all, and two, when I actually added up the amount of time spent doing it based on the amount of compost I made in return, I was shocked by how costly my own compost was. If I spent a total of five hours turning a pile of compost on a few different occasions, it might yield around one yard of compost. For $40, I can buy a yard of compost from a local fellow who makes it far better than I can, plus I don't have to strain my back turning it by hand. Do a true cost-benefit analysis of everything you do; see if there's a better way to do it, or cut it out entirely.

Apply a questioning attitude to everything you do on the farm. If you're spending a lot of time pulling weeds, ask yourself: Why am I pulling weeds all the time? Is there a way to avoid weeds altogether? Is there something I can do before weeds become a problem? The same can be said for crop production. I used to have to thin carrot plantings by hand. This is a huge waste of time, because unless you can sell those thinnings, you're not getting paid for that labor. So, for me the solution was to seed carrots in a way that didn't require thinning. This meant investing in better seeding equipment, like the Jang Seeder.

Another aspect of working smarter is to stop being a perfectionist. If you seek perfection in everything, you will rarely be satisfied. If you can achieve 85% quality, that's good enough, and most people can't tell the difference between 85% and 100%. Imagine filling up a glass of water under the tap. The first 85% fills up very quickly, then to get it to full, you have to slow the tap down so it doesn't overflow. This is true in life, and especially many tasks on the farm. Taking the extra time to get perfectly straight rows when you're direct seeding isn't that important, because it doesn't make any noticeable difference in your production. Most people comment on how straight my rows are anyway. Do the job right, be effective, but don't kill yourself by making everything perfect.

Can time do some of your work for you? Covering a lawn, using tarps over time to prepare a new plot, can kill most of the grass. This means less work and resources you have to expend to get the job done. Using the stale seedbed technique to prepare beds, then tarping them for a week or two, is a time-based technique, using appropriate technology. Let time do the work of getting weeds to germinate on the soil surface so you can flame weed or simply let the tarps smother them. This saves massive amounts of time because having no weeds in your beds means you don't have to spend time picking them out. When I harvest carrots, I immediately soak them in water for the day. Soaking makes washing carrots very easy because the time they spend in water loosens up the dirt, making them a lot easier to clean. For every task that you do, think of how time, technique and technology can expedite tasks and make jobs easier for you.

### The 48-Hour (or Less) Farm Week

When I tell farmers that I work an average of 48 hours a week, and often less during the summer, they don't believe me. How is it possible that a farmer works less than most small business owners? Well, there are many reasons. The most notable comes from my crop selection. I grow only a handful of summer crops such as tomatoes, pattypans, peppers and eggplants. Many farmers are their most busy during the summer months because so many of their crops are ripening at that time, so there is a lot of work to be done, harvesting and keeping up with weeds. Weeds are hardly an issue on my farm because of proactive techniques I will outline further in this part. Also because I live close to my farmers market, I need to wake up on Saturday only 30 minutes before driving down that morning.

One strategy that saves me a lot of time is I balance out my weekly harvesting. Instead of harvesting everything on one day for Friday restaurant deliveries and the Saturday market, I balance it out throughout the week. Generally, I try to harvest the least perishable crops (for example, root crops) near the beginning of the week, then the most perishable (greens and microgreens) near the end. I have noticed over the years that I am most productive when I'm well rested and not doing the same task for too long. My most productive hours of the day are the first few of the morning, and the last hour or two after a siesta during summer. When you come at something fresh, you will approach it with a lot more vigor. This is partly the idea behind spreading out the harvest. Also find the optimal time to harvest a particular crop so that it will be less work to harvest or wash. For example, if weather during the week is perfect, I will usually harvest salad greens on a Thursday, so that Friday can be spent assembling orders and portioning. However, if the weather is not perfect and it's rainy on harvest day, I have to wash those greens, which can add another four hours to the week. I watch the weather forecast every week and plan to harvest greens on the best day. If that means Tuesday is going to be the driest day of the week, I'll harvest my greens then. If I don't have to wash the greens, they will have a much longer shelf life. I find in general that it's best to avoid harvesting anything in the rain if you can. If your boots are getting stuck in mud, that's slowing you down. If you're stopping every now and then to change your shirt, socks or gloves, that's slowing you down. Sometimes it's inevitable, but as often as I can, I will structure my harvest based around the weekly weather.

### Structuring the Week

There are many tasks on the farm that have to be done on a regular basis. These tasks include planting, harvesting, nursery work, turning over beds, new plot preparation, moving around poly low tunnels, flame weeding and opening and closing up greenhouses.

Managing multiple plots can be hectic, so having a strategy in which you balance out the workload is very important. For plots in my Hi-Rotation areas, working on these is simple, because they are so close to one another. However, some plots in my Bi-Rotation areas are a little far away, so I like to accomplish as many tasks there as possible in one trip. Whether I have any planting or harvesting to do there during the week, I'll combine all the tasks at that plot in one trip. For these trips, I'll use my truck and pack up my planting equipment, emergency irrigation tools and harvesting equipment. I'll get everything that needs to be done on that plot at once.

### Weekly Routines

The setup period is the busiest time of our season. From mid-March until mid-May, I work around 60 hours a week. This is a busy time for most farmers, as you're getting all the needed infrastructure together for another season: setting up irrigation, cleaning out greenhouses, putting up poly low tunnels, preparing new plots, starting seeds in the nursery and preparing large areas of beds for stale seedbeds. Life on the farm is pretty hectic at this part of the year, but once infrastructure is all set up, it's more or less smooth sailing for the rest of the growing season.

#### *Quick Overview*

On a weekly basis, my farm serves nine major customers and one farmers market. Seven of those customers are medium- to high-end restaurants of varying sizes, and two of them are organic produce distributors. I deliver to the restaurants every Friday by 2 PM, to one of the distributors by 9 AM on Tuesdays and the other by Wednesday afternoon. Distributor #2 sometimes picks his orders up on Wednesday. I receive an order from distributor #1 on Sunday afternoon, so that means I can start harvesting for that customer on Monday morning. I receive distributor #2's order late Monday afternoon. For the most part, I can start harvesting for both distributors around the same time. I do this so that I can harvest from the same locations at the same time, which is far more efficient for me because my farm is multi-locational.

I receive all my restaurant orders by Thursday at 9 AM, and that means that the bulk of the harvesting and washing for restaurants is done on Thursday, and most of the processing, washing, portioning and packing is done on Friday. In the summer months, when the volumes are high and the temperature is too, we'll start harvesting on Wednesday. Early harvest items will be beets, carrots, tomatoes and other less perishable crops. My strategy is to stagger the harvesting period over a few days, so that we're harvesting at the optimal times of the day. In the summer, that's in the early morning and evening.

## Preventive Weeding

Weeding has got to be the one task that I want to do least. I actually enjoy pulling weeds and love the sensation of pulling them out of the ground. That's the problem. It's something that I'll do, and hours will pass by, then I'll be wondering where the day went. The best way to weed is to not weed at all. Taking proactive and preventive steps to reduce weeds will be your best time-saver on the farm.

If you ask most organic farmers, they'll tell you that a huge portion of their labor is dedicated to weeding. This is part of the reason that a lot of farmers don't end up with much profit at the end of the season. Weeding is one of those tasks that doesn't have an easily measurable outcome. You can't count it or put a price on it, as you can with pounds of greens or bunches of radishes. That's why you must not waste too much time on it.

### Preventive Weeding for Beds

Bed weed prevention is done through the *stale seedbed technique*: we prepare the beds, weeks prior to planting, to allow weed seeds to germinate; then we eliminate those weeds either through smothering or flame weeding. At the beginning of the season, I prepare large areas of beds using this technique. I begin preparation either through rototilling or using a no-till technique with a broadfork and tilther. With either of these approaches, I am adding all of my compost and fertilizers to the bed, then working the amendments into the soil. Once that is finished, the beds are watered heavily, then covered with a tarp.

The idea is to create an optimal environment for the weeds in the top layer of soil to germinate. The length of time this takes will vary for each season. At the beginning, it takes at least two weeks, but in summer, germination may only take a week. Once the weeds germinate, you have two options: (1) either keep the tarp on for another two weeks (if you have the time for that), or (2) if you need to get another crop in there right away, bring in the flame weeder and torch the beds. A flame weeder burns only the top 1⁄32 inch of soil and does not hurt your soil structure.[1] But the flames do destroy all of those new weeds. Now, when you plant, you're going to seed or transplant directly into the soil without disturbing it to bring up any remaining weed seeds. Once we are into steady turnover of beds by mid-season, we still use the flame weeder. Once a crop has been harvested and the remaining crop residue has been removed, it's time to replant it. We pull any large weeds out of the bed by hand, then flame any small weeds that have recently germinated, if necessary. From here, we turn the bed over with a no-till technique to prepare for replanting. If there aren't a lot of external reasons for weed seeds blowing onto your land, your beds will get less weedy over time through this process.

FIGURE 42. Flame Weeding with a Five-Torch Weeder.

#### *Preventive Weeding for Perimeters*

I use heavy landscape fabric or some kind of mulch, usually wood chips in perimeter areas. My default is to use landscape fabric because it's easy for one person to lay out a perimeter in 30 minutes, whereas shovelling wood chips can take all day. The only time I will use wood chips is if I have long-term tenure on that piece of land. If it's a plot that I may be at only for a few years, then landscape fabric is the best approach. See photos #6 and #12 in the photo insert section.

#### *When and When Not to Weed*

Knowing when to weed is also very important. Spending time weeding a bed of radishes is not an effective use of time; because radishes grow so fast that they will compete well with the weeds, and then they'll be out of the ground.

However, it is important to always make sure that you pull weeds that are going to go to seed. One weed will produce thousands of seeds, so you need to stay on top of this, especially around crops like any loose-leaf greens. These crops will get multiple cuts, so having the beds around them weed free is very important. Otherwise you will spend too much time sorting out the crop from the weeds once it's harvested.

When you're planting crops for the longer term, even in the case of carrots, establishing a stale seedbed is very, very important, but there will inevitably be times when you have to get in there and pull some weeds.

### Cutting Your Losses

"You've got to know when to fold 'em, know when to hold 'em," as the classic Kenny Rogers's tune goes. It's very true in farming. You need to know when to cut your losses, correct and move on. On the micro land base of an urban farm, it is absolutely critical to maximize your production. Let's face it, in farming sometimes the unexpected happens. It could be extreme weather, an irrigation failure, some bad seed—there are many reasons why a crop could fail. The trick is to identify the problem quickly, correct it and start over.

For example, you might have just planted a crop of arugula and noticed that only 70% of it germinated. First, see if there were external reasons to why it germinated poorly. Was there a problem with irrigation? Was it a certain part of the bed that didn't germinate? (If that was the case, it may be a watering issue.) Did some pest like an insect or bird eat the newly germinated crop? Once you start going down the line of possibilities, you should be able to identify how the problem occurred. Once you do know, you need to take immediate steps to fix it, and then move on.

If I see a bed or crop has suffered more than 30% damage and will only yield 70%, I may decide to keep it in and ride the loss out, expecting a lower yield. If I estimate anything beyond a 30% loss (if it's a quick-growing crop), I will reseed right away and except a slightly delayed harvest. That's what's nice about quick-growing crops and why I keep them to my Hi-Rotation areas. I want to see if there are any problems fast, so I can react quickly.

Sometimes I might not be selling as much of a crop as I thought I would, and in order to cease losing money on a crop I'm not selling, I may pull it out of the ground and plant something else that is in more demand. During one season, I had an area of kale planted. There were six beds of it in production on one plot. I planted six beds with the plan of harvesting two per week, and after the last two were harvested, the first two had regenerated to be picked again. This worked well for the first two cycles, but by the time I went to pick the last two beds, the first two were already fully regenerated. At this point I was producing too much kale—or more than I could sell, at least. I was planning to put in fall carrots here after the kale on August 1. So, I decided to pull out two beds July 1, and plant two beds of baby beet greens in their place. It wasn't ideal, because I was planting a Quick Crop in a Bi-Rotation area, but it was for a short period and didn't interrupt my workflow. During summer, beet greens are a 21-day crop and I'll get a couple of cuts before turning them under. I was still able to plant my carrots on schedule because the beet greens had such a short cycle.

# Harvesting

Very few tasks on the farm happen as often as the harvest. I strive to have my farm in a constant state of production by focusing on crops that are constantly being replanted and harvested. The best way to achieve steady cash flow on a farm is to constantly have product to sell. Once we're in our main season, we are harvesting something every day. Harvesting is a task that needs to be as streamlined and efficient as possible.

There are a number of harvesting protocols for the various crops on the farm, and there are some core principles that apply to all of them, regardless of what we are harvesting. Be efficient with the movements of your body. Make the least amount of back and forth movements with your hands or tools as possible and maximize the amount of product you are holding, cutting or pulling. For example, when harvesting greens by hand, you want to build up as many greens in your hand as you can when you're cutting—before you move your hand to dump them into your basket. If you pick one leaf at a time and move your hands back and forth for small amounts, you will take much longer to finish your harvest. When bunching radishes, make a whole bunch in your hand before you put any down, and every time you move your hands to your harvest bin drop one complete bunch.

Efficient movement also applies when walking back and forth from your place of harvest to your harvest vehicle or wherever you are taking the product. Bring something with you every time you go from point A to point B. Bring as many harvest tools as you can comfortably carry when you are heading out to the field; this minimizes the amount of times you have to walk away from the plot to get another tool or bin that you need.

Harvesting from multi-plot farms can be a lot to manage logistically because you need to carefully assess what you're going to be harvesting that day, where the crops are and what tools and equipment you're going to need to get the job done quickly and correctly. When I'm packing up for a harvest day, I load my bike trailers or truck with everything I will need, and sometimes a few extra tools just in case. I will often bring a stirrup hoe and pitchfork with me in case I see any potentially unruly areas with weeds coming on. I will also bring a small bag with a few irrigation line couplers, tools and extra timer batteries just in case I find any problems with a sprinkler system when I go to a plot. Not having these with me has cost me a few trips back and forth to plots, when I should have just been focusing on the harvest.

### Harvest Tools and Gear

- Harvest tally sheet on a clipboard—to keep track of the harvest, make any notes or modifications and track my yields.
- Analog scale—to weigh the harvest as we go
- Painter's tape and marker—to label bins as we harvest
- Harvest bins and large totes
- Pitchfork—to harvest carrots and pull deep weeds or invasive grass if needed
- Harvest knife—to harvest spinach and bunched herbs
- Quick Cut Greens Harvester with four spare batteries
- Mini bin of elastic bands—for bunching herbs and roots

## The Seven Stages of Harvesting

We follow seven stages to achieve a successful harvest each week. As long as we have all the necessary tools and equipment we need for the harvest day, this is how we go about getting everything we need harvested and ready for deliveries and the market.

1. Structure the harvest week
2. Gather your orders and compose a harvest sheet
3. Harvest at the right time of day (dry and cool)
4. Decide in advance the order in which you harvest
5. Place your harvest in the shade and remove the field heat
6. Track your harvest as you go
7. Place your harvest in cold storage

### *#1: Structure the Harvest Week*

At this stage, we are looking at what crops are being harvested during the week, where they are, and which days will be best to harvest them. Quickly checking the weekly weather forecast might change your plans slightly. During the early and late parts of the season, all the harvesting and processing on my farm is done on Thursday and Friday. By summer, once the farm is into full production we need at least three days to get all the harvesting, washing and packing completed. On Wednesday, we'll start with crops that are the least perishable (like beets, carrots and tomatoes) and/or crops that are the farthest away in the network of farm plots. Most of those will be harvested and/or washed by this first harvest day. If some farm plots are farther away in the network chain, we will harvest all of the crops we need from that particular site. For example, if we're harvesting beets and carrots from a site that also has kale it makes sense for us to harvest kale as well, so that we don't have to come back the next day.

In general, the most perishable crops will be harvested the second day. Thursdays is when we harvest most of the baby roots like radishes, turnips and greens, then microgreens at the end of the day. In high summer, Friday is when we pack and portion everything. To make the best use of our time, we will mix harvesting with other farm tasks. For example, if we had to go to one of our more distant sites for any particular reason—whether to fix irrigation, plant something or move poly tunnels—if there was something to be harvested at that site, we would try to harvest then as well.

### *#2: Gather Your Orders and Compose a Harvest Sheet*

Before we can start harvesting, we first need to gather our orders and break them down into a harvest sheet that we will take into the field. This is done by adding up the totals of each type of crop to be harvested, then making a list of all of those totals to be harvested.

#### Farmers Markets

What I take to the farmers market each week is loosely based on what was brought last week, then slightly modified based on sales or what was left over. For example, if I brought 100 bags of spring mix one week, but only sold 80, I will bring around 80 the next time around, and try to move any extra volume that would have gone to the market to some other customers, like chefs.

#### Landowners

Our landowners receive an e-mail on Monday that lists all of the product we have available for them. They can choose $20–$30 worth, depending on the season, and they must have those orders back to us by Thursday morning. These quantities get included in the harvest sheet.

#### Restaurants

Our weekly fresh sheet goes out to chefs every Monday morning, and we start to receive orders from them on Wednesday. The cutoff for restaurant orders is Thursday morning at 9 AM. I input all of my restaurant orders into a spreadsheet as I receive them. If I haven't received everyone's order by 9 AM, I'll follow up with a text message, phone call or e-mail. If I still don't hear back from them, I will head out to the field and begin harvesting anyway. For customers that are very consistent, I will just cut and paste their last week's order into the harvest sheet and begin harvesting based on that. During the months of summer, we start harvesting on Wednesday, and I am speculating some orders based on what they had the last week. Tracking everything on spreadsheets allows you to learn and predict regular customers' orders. You will get to know and memorize what your main customers will need each week. It will take some time, but over the years, I've been able to predict my customers' orders based on previous seasons.

#### #3: Harvest at the Right Time of Day

Harvesting when the weather is cool is better for the shelf life and appearance of the product, and it's also better on the people doing the work. Some crops are more susceptible to heat than others, but as a general rule it's better to harvest during the cool times of the day. Two exceptions to this are root and summer crops. If you have to, they can be harvested in the heat, but some protocols must be followed to ensure their quality. See Chapter 33 and Part 10 for more details.

During spring and fall, the time of day isn't quite as important because temperatures are generally a lot cooler. What's more important during these seasons is to harvest when it's dry. In the case of greens, this will reduce the amount of processing time, but it's also not very enjoyable to work in the rain.

If you must work in the rain, do as many tasks as you need to get done that particular day in greenhouses or under cover: that could be processing some crops that were already harvested or harvesting in the greenhouses. This change of plan might buy you enough time to wait out the rain. I constantly check the weather forecasts to see what it will be, and I will plan my harvest week around that. If the weather is agreeable and steady, I will follow the regular harvest protocols laid out in Stage 1.

#### #4: Decide in Advance the Order in Which You Harvest

The order in which you harvest crops is very important: harvest the least perishable crops first and the most perishable last. Because we are farming on multiple plots, there are some logistics to take into account. When we arrive at a plot where radishes, herbs, turnips and greens need to be harvested, the first thing I harvest is the radishes and turnips, followed by the greens and herbs. Leave the most perishable crops until the end of the harvesting time so to minimize the amount of time between being harvested and being placed in your cooler.

#### #5: Place Your Harvest in the Shade and Remove the Field Heat

If it's sunny, after you have harvested a basket, bin or tote full of product, immediately put your bin into a shady area. If you're starting with root crops, after you've harvested a bin's worth, spray some water on them while they sit in the shade. Sometimes, I will also fill the tote with water, and let the roots sit in it while I harvest more product. This has a twofold effect: first it removes field heat from the veggies, and it also makes the dirt on them come off easier when it's time to wash. Just make sure to drain the totes out before you pack them into your vehicle. With summer crops like tomatoes, pattypans and peppers, just put them in the shade.

Generally speaking, I never put water on greens after they have been harvested. The only time I may add a splash of water is if they were harvested when the temperature was warmer than what's ideal. Put greens in a shady area immediately, and keep the lid on the tote so that they don't lose moisture.

### *#6: Track Your Harvest As You Go*

Every time I fill up a tote with either greens or root veggies, I weigh it or count it and then mark that information onto my harvest sheet (see Figure 9). You need to keep track of what you've harvested as you go, so that you'll know what else you need. Then I take painter's tape and marker and label the tote so that when it goes into the cooler I can see what it contains. As you're harvesting, it's important to track how much your plots are yielding. If you cut a bed of arugula, write down how much you harvested. If you cropped out half a bed of carrots, write that down: note how much of the bed was harvested and the weight. All of this information will help future farm planning because you'll know what to expect from certain crops.

### *#7: Place Your Harvest In Cold Storage*

Once you arrive back at your home base, immediately put everything into your walk-in coolers. I find it's helpful to put things that don't need to be processed near the back or bottom, so that products you need first are near the top or front. This will save you shuffling totes around unnecessarily.

## Protocols

Here are some specific protocols for crop varieties we grow on my farm. For comprehensive detail on each crop, see Part 10.

### *Greens*

Greens (baby leafy greens like lettuce, arugula, mustard greens and spinach) can be the fastest or slowest crops to harvest, depending on the weather. Checking the forecast can mean the difference between a quick harvest (done in two hours) or a miserable slog in the rain that takes almost all day, or even longer when you factor in washing time. If the weather is perfect, we will harvest all of our greens on a Thursday in preparation for Friday restaurant deliveries and a Saturday farmers market. But in the event of a bad weather forecast, it is a good idea to harvest the greens on any day you can avoid harvesting in the rain or when the greens are wet. I set the irrigation timers on the farm around when we usually harvest greens, so that when we go to harvest in the morning, the greens areas are dry. If we're planning to harvest on Thursday morning, I will set the timers to not water on that day, and as we leave the plot after the harvest, I will turn them on manual to water right then. This is mostly a concern for the hot summer, where we need to water each day. In all other seasons, it's not as important.

Make sure your sprinklers are set in such a way that they don't cause soil to splash up onto the greens. This saves us a huge amount of washing and processing time. See photos #19 through #23 in the photo insert section.

### *Root Crops*

We use a slightly different method to harvest each root crop. Carrots and scallions follow the same protocol: use a pitchfork to loosen up the soil, and then pull from the ground. With carrots, we twist the tops off and throw them in a bin for the compost as we harvest. Or for carrots with really long greens, we cut the tops off with a very sharp knife while they are still in the ground (keeping an inch or two of green on them), then fork from there.

Radishes and turnips are harvested in the same way as each other, as they most often all mature at the same time, so they can all be harvested at once (*cropped out*). There are some times during the season that a bed of radishes or turnips will be harvested over two weeks: thin harvested on the first week, then cropped out on the second.

Beets are most often thin harvested, meaning that we are picking the mature ones first, and leaving the rest for another week. When harvesting root crops during hot days, after you fill a bin take it into a shady spot on your plot and spray it down with some water, or in the case of carrots and scallions fill the tote up with water. This will immediately remove the field heat, and make the veggies easier to clean when it's time to process.

See root harvesting photos #24, 25 and 26 in the photo insert section.

### *Summer Crops*

Harvesting summer crops such as pattypans, zucchini, tomatoes, peppers and eggplants is very simple. Not as much care has to be put into choosing the right time of day for harvesting summer crops, as they don't suffer from the same amount of deterioration after picking as greens or even root crops do. Tomatoes don't even need to be placed in the cooler. As long as your tomatoes are in a shady and ventilated area, they will keep just fine. Pattypan squash are picked by hand, and we wear long sleeves, pants and gloves when picking them as the plants have small thorns that can cause itchy skin. When harvesting zucchini, we use a small knife or trimming scissors. When tearing zucchini off by hand, they sometimes will break at the base, so making a small cut will sever it cleanly at the stem. Pattypans and zucchinis both should be placed in the cooler right after harvesting. Peppers are picked by hand and should be refrigerated after harvesting.

# Post-Harvest Processing

*Processing* refers to washing, sorting and cleaning vegetables to be packed for either a market or delivery to a restaurant. If someone had showed me the techniques I developed and describe in this part early in my farming career, they would have saved me a lot of time. Hopefully my experience can be valuable to you.

The stages after the harvest are as follows:

1. Remove field heat
2. Wash, spin and/or dry
3. Portion and/or pack

These stages do not always follow this sequence, but in most cases processing will begin and end with stages 1 and 3. Tomatoes do not need to be cooled or washed, but they will be portioned and packed. Not all vegetables need to be washed either. Peppers, eggplant and summer squash don't need to be washed, but they should be cooled and portioned.

The key to getting top prices for your products is to have them looking near perfect. You want to showcase the color and freshness of vegetables to your customers, so that they will keep coming back for more. Being efficient is equally important. Try to establish a rhythm to your workflow, and keep things moving along consistently.

Three types of vegetables are common on our farm and involve the most processing technique to prepare them for sale.

## Greens and Microgreens

Washing greens is something I do only if I absolutely have to. Washing considerably reduces the shelf life of fresh greens. They keep far longer and display better if they are not washed. For the most part of the season, I don't have to wash greens very often. My irrigation is set up so that it doesn't splash dirt up on the greens; they are usually very clean right from the field. However, during rainy times in the season, it is inevitable that the greens will be splashed with dirt. After greens have been harvested, they will sit in the cooler for at least a few hours. This removes the field heat; it is very important that this happens before they get washed. If the greens are harvested in warm weather, splash a little cold water on them. It is good to let greens sit in the cooler for a while. This causes the leaves to firm up.

The three stages for processing greens and microgreens are

1. Wash
2. Spin
3. Dry and sort

In some cases, you may not need to do all three. If your greens are just a little bit damp from spring morning dew and don't have any standing water on them at all, you can just go to stage 3 (drying and sorting). If you have had a light rain and the greens are a little bit wet but not dirty, then you may be able to skip stage 1 and just spin, dry and sort. If, however, your greens have been getting rained on for days, they will be very wet and gritty with dirt that splashes up from the ground. They need to be washed and go through all three stages. With sun shoots and radish shoots, you will always follow the three stages.

### *Stage #1: Wash*

I will have three totes in front of me, on the surface of my washing table. The tote on my left came from the cooler; it holds our product to be washed. The tote in the middle is a medium-sized tote ⅔ filled with clean, cool water. The third tote on my right is a large tote with holes drilled into the bottom for drainage. I hang one laundry bag inside this tote and use a couple of spring clamps to hold it wide open. I put the greens into this bag when I have washed them.

To wash, I take large handfuls of greens from tote #1 and dump them into tote #2. I swish them around a couple of times, and then take that handful of now-washed greens and dump them into tote #3. I repeat that process until either the laundry bag is full or tote #1 is empty. Over the course of washing a couple of bins, you may need to empty bin #2 and fill it with fresh water. I just dump dirty water onto the washing table and it drains through. This is why I do this process on the washing table.

When washing sun or radish shoots, as you dump handfuls into your washing tote, flick your hands along the surface of the water; by splashing away from you, you will notice that the seed hulls will build up on that side of the tote. Now, if you grab the shoots under the water and pull them towards you, away from the hulls, you will have a handful of shoots with very few hulls, and you can then dump them into tote #3. Every so often, I will use a small colander to scoop the hulls off the surface and throw them in the compost. See photo #27 (washing shoots) in the photo insert section.

### *Stage #2: Spin*

From here, the laundry bag full of greens from tote #3 goes into the spinner. Make sure the bag is evenly distributed around the circumference of the spinner, so the weight is spread around. If not, the machine will make a lot of noise as it moves around. I set the machine for spin cycle, and let it run. While that's running, I will continue to wash more greens and fill up another laundry bag. Once that one is full, the first spin cycle will usually be finished. See photo #28 (spinning machine) in photo insert section.

#### Stage #3: Dry and Sort

I pull the bag of freshly spun greens out of the spinner and dump it onto the drying screen. I spread those greens evenly over the screen, and then run the fans at full. Once I have all three stages going at once, I've got a bit of an assembly line going. I'm washing, spinning and drying all at the same time. When greens are drying on the screen, I will gently shuffle them around a little bit to expose more leaf surface to the air, and I will also use this opportunity to pick out any bad leaves, weeds or critters I may find. Having the greens on the screen really spreads them out so you can see what needs to be removed.

When drying shoots, the last remaining hulls will fall through the drying screen as you move them around. You have to be careful not to dry shoots too much; otherwise they wilt. See photo #29 (sorting on the drying table) in the photo insert section.

### Bunched Root Crops

We don't always bunch our root crops, but when we do it's done this way. Sometimes the greens are removed in the field, sometimes I remove them at the processing stage and sometimes I don't remove them at all. It all depends on what your customers want. When bunched root veggies come from the field, they can either go into the cooler for a while or they can get washed right away. If they are going to sit in the cooler for a while, I like to give them a little spray of water before. I find it makes them easier to clean later.

If you are going to remove the greens, there are two stages to processing root crops:

1. Remove greens
2. Wash

#### Stage #1: Remove Greens

Sometimes you can remove greens from radishes and turnips in the field. However, there are times we will remove them during processing because

- a customer has requested it
- we may not have had room to leave the greens at the plot
- we wanted to harvest them quick for reasons like bad weather or rushing an order

To remove greens from roots, I use a good pair of scissors to cut the greens off. On my washing table, I will put a large, perforated tote of roots on my right side on the table, a large tote on the ground at my feet and a large, empty tote on the left side of the table. I will grab a bunch of roots with my left hand (I am right-handed), cut the greens off about an inch above the elastic band, those greens will fall into the tote at my feet and the bunch of roots goes into the tote on the left. This process takes me between two and four seconds per bunch.

#### *Stage #2: Wash*

To wash the bunched root vegetables, move a tote full of bunched roots to the right side of the table and put a clean tote on the left. Turn the water sprayer on a sharp spray, and keep it running. Holding the sprayer in my right hand, I grab the bunch of roots with my left hand and spray it clean. I will spin the bunch around 360° so that I get water everywhere. I will basically spray every part of the bunch clean, including the greens. Once you get into a rhythm, washing will only take about five seconds per bunch. After you clean your bunch, it goes into a clean bin. When I have a large restaurant order—like 30 or more bunches—I will just wash them and put them directly into a waxed box for that order. It is also possible to harvest and wash radishes, turnips and beets in a similar way to carrots, by removing the greens as you go and not bunching them. This would be done for bulk orders when selling by the pound.

### Scallions

Scallions (also called spring onions) are the most tedious vegetable to wash; they take more time than most other root veggies. This is because their roots hold a lot of dirt from when you pull them in the field. One way to make this processing easier is to stand freshly harvested scallions upright in a tote, and fill it up with a few inches of water. If you can let them sit over night, or for at least a few hours, they will be far easier to wash.

Washing scallions is divided into four stages:

1. Spread them out on the table and spray them down
2. Pull off some of the dead skin
3. Trim roots
4. Spray down again, bunch or pack

#### *Stage #1: Spread Out and Superficially Spray Down*

Take as many scallions from your tote as you can hold and spread them out on your washing table. Give them a good spray down. You're just trying to get most of the heavy dirt off.

#### *Stage #2: Pull Off the Dead Skin*

Take a small handful, almost the equivalent of a four-ounce bunch, and fan them out in your hand. You're going to hold onto them near the middle of the length of the scallions. Now you're going to pull off any dead skin, just generally cleaning them up. Complete a large pile then go onto the next stage. Don't complete an entire bunch at a time because you will be picking up and putting down scissors many times over, and this time adds up.

#### *Stage #3: Trim Roots*

Once you've removed the dead skin, use scissors to trim the roots down to about

¼ inch or less. Start to make a small pile on the left side of the wash table.

### *Stage #4: Spray Down and Bunch*

Once you've got a good pile of peeled and trimmed scallions, spray them down thoroughly. Shift them around a bit as you're doing this to get all the remaining dirt off. They will clean a lot easier at this point because there is very little left (like roots and excess skin) on them to hold dirt. Once you've sprayed that pile of scallions, now you can bunch them or pack them into bulk orders.

## Loose Roots

Washing loose root vegetables (beets, carrots, radishes, turnips) is considerably faster and more thorough. I do bunches only on specific request. After I harvest, I will often soak the crop in water in the tote and let them sit there while I harvest other veggies. If I'm going to process them later in the day, when I take them back to the home base I will fill the tote again with water, and let them sit in the shade, next to my wash table. If I'm washing them the next day, I will just put them into the cooler right away.

There is one stage for washing with this method, but there can be a second stage if you need to sort the root veggies by size. The first stage can be very fast. With this method, I can wash 100 pounds of roots in 30 minutes. All you will need is a perforated harvest bin for washing and a clean one to hold the washed product.

### *Stage #1: Wash the Roots*

From the tote from the field, dump about a two-inch-deep layer of roots into your washing tote. Turn on the water sprayer to nearly full blast, hold the sprayer with your right hand and spray into the tote. As you're spraying, move the tote back and forth with your left hand. This will shuffle them around so that they all get sprayed equally. It should only take a couple of minutes to wash a whole tote. Once the roots are totally clean, dump them into a clean tote on your left. You will keep doing this process until you fill a tote to the top. Leave enough room so that you can stack another tote on top without damaging the product in the tote below. I keep doing this and stacking up the totes until everything is washed. I like to let them sit for around 20 minutes so that they drip dry before they go into the cooler. See photo #30 (washing roots) in the photo insert section.

### *Stage #2 (Optional): Sorting*

I don't sort unless the sizes vary a lot. Generally, I try to grow small- to medium-sized root veggies because I get a premium price for them. Once they have all been washed (stage #1), I will put them on the center of the washing table and put clean totes on the left and right of it. Now, I just go through and throw smaller ones into the left tote and larger ones into the right.

# Portioning and Packing

After all of your product has been harvested and washed, now it's time to pack it and take it to your customers. On my farm, I have a small team take care of packing for the market as it involves tedious labor of portioning hundreds of items into small, $3–$5 units. Packing for restaurant orders is a lot different; the unit volumes are a lot larger. The work is a lot less tedious, and usually one person can handle all of it. One order may be anywhere between $200–$1,000 in value, comprised of many different items. Extra care must be taken to pack orders for chefs. When customers are spending this much money on a weekly basis, you need to get it right.

## Packing for Market

On my farm, all of the packing for market is done on Fridays. The harvesting and washing has all been done prior to this. I work with a small crew of people on Fridays. Packing for market is very easy and non-strenuous work, and I found it very easy to find people who are happy to do a couple of hours work in exchange for vegetables. When we are packing, we're working in a sheltered area with lots of shade. Each packer stands at a table and works on one product at a time. They will have a tote of finished product that came from the cooler, usually on their left side, with a scale in front of them, with twist ties, bags and/or elastic bands. On their right, they will have another tote to hold the portioned product. As they fill a tote with bags or bunches, they will label the tote with painter's tape and a marker, recording what the item is and how many are in the box. Then each full tote goes into the cooler at a specified place. When we're in full production during high season, we're running two coolers. One holds product to be portioned, and one holds portioned product and assembled restaurant orders.

Because some of our root crops for market are bunched in the field, all we are packing for market on Fridays are greens, microgreens and loose carrots; we are also bunching kale, chard and scallions. Items like tomatoes, pattypan squash or peppers get assembled and weighed into berry baskets at the market in the morning as we set up.

The chart below outlines all the items we take to farmers market, the weight, how they're packed and the price at which they are sold. We actually sell all items shown at the price of $2.50, for $3 each or in a two-for-$5 mix-and-match deal, but for ease of listing and accounting, they are listed as $2.50. All items are sold as units except for large heirloom tomatoes, which need to be sold by the pound because no two tomatoes are the same size. For this case, we bring a small scale to the market to weigh items on the spot.

When we are packing bags for market, we follow a rule of 85%. When the greens are on the scale, if you are within 85% of the needed weight, that's good enough. It wastes valuable time when a person takes extra time to make each bag exactly perfect. Over hundreds and thousands of bags, that combined time can really add up. If we're weighing four-ounce bags and you're at 4.4 ounces, the difference is negligible to most customers. Generally, we'll try to be slightly over rather than under weight.

### *Packing for Chefs*

Restaurant orders are high volume, and we use different types of bags and boxes. I collect recycled waxed boxes from organic grocery stores and use those for packing restaurant orders. These boxes get labeled by destination, with painter's tape and markers. I also collect shallow tomato and avocado boxes for packing case lots of tomatoes; you can find these at the back of organic grocery stores as well.

Greens are packed into case-lot-size bags that weigh two pounds. These bags are supplied on a roll and can hold two pounds without having to pack the produce in too tightly. For very large greens orders, I will use bags similar to garbage bags to fill ten pounds in a box. Make sure you're using something that is food safe. Items like tomatoes are packed into flat three-inch-high tomato boxes and are delivered as a ten-pound case lot. Kale and chard are sold as a five-pound box. These greens are packed into waxed boxes, and we will put down a plastic lining in the box before we pack it.

**FIGURE 44.** Packing Chart for Market.

| Product | Type | Market Price | Unit Weight | Packaging | Item Details |
|---|---|---|---|---|---|
| Arugula | salad greens | $3/two for $5 | 4 oz bag | 5" × 3" × 11.5" bag w/twist tie | |
| Arugula (large bag) | salad greens | $5 | 9 oz bag | 10.5" × 15" roll bag w/twist tie | |
| Basil | herbs | $3/two for $5 | 4 oz bag | 5" × 3" × 11.5" bag w/twist tie | |
| Beets | root veg | $3 | 3-4 beets | bunched with elastic band | sold with tops |
| Bok Choy | greens | $3/two for $5 | 8 oz bag | 10.5" × 15" roll bag w/twist tie | |
| Braising mix | salad greens | $3/two for $5 | 4 oz bag | 5" × 3" × 11.5" bag w/twist tie | |
| Carrots | root veg | $3 | .75 lb. bag | 5" × 3" × 11.5" bag w/twist tie | topped and bagged |
| Cilantro | herbs | $2 | 2 oz bunch | bunched with elastic band | bunched in the field |
| Dill | herbs | $2 | 2 oz bunch | bunched with elastic band | bunched in the field |
| Eggplant (baby varieties) | fruits | $3/two for $5 | 10 oz basket | pint size berry baskets | |
| Kale | greens | $3/two for $5 | 8 oz bunch | bunched with elastic band | |
| Parsley | herbs | $2 | 2 oz bunch | bunched with elastic band | bunched in the field w/elastic band |
| Pea shoots | microgreens | $3/two for $5 | 2 oz bag | 5" × 8" bag w/twist tie | |
| Pea shoots (large bag) | microgreens | $5 | 5 oz bag | 4" × 2" × 10" bag w/twist tie | |
| Radishes | root veg | $3/two for $5 | 8 oz bunch | bunched with elastic band | tops removed after bunched |
| Radish shoots | microgreens | $3/two for $5 | 2 oz bag | 5" × 8" bag w/twist tie | |
| Red Russian kale | salad greens | $3/two for $5 | 4 oz bag | 5" × 3" × 11.5" bag w/twist tie | |
| Salad mix | salad greens | $3/two for $5 | 4 oz bag | 5" × 3" × 11.5" bag w/twist tie | |
| Salad mix (large bag) | salad greens | $5 | 9 oz bag | 10.5" × 15" roll bag w/twist tie | |
| Salad turnips | root veg | $3 | 8 oz bunch | bunched with elastic band | |
| Spicy salad mix | salad greens | $3/two for $5 | 4 oz bag | 5" × 3" × 11.5" bag w/twist tie | |
| Spring Onions | root veg | $2 | 4 oz bunch | bunched with elastic band | |
| Spinach (baby) | salad greens | $3/two for $5 | 4 oz bag | 5" × 3" × 11.5" bag w/twist tie | first cut, premium salad green |
| Spinach (large leaf) | greens | $5 | 12 oz bag | 10.5" × 15" roll bag w/twist tie | second cut and on, for cooking and juicing |
| Summer squash | fruits | $3/two for $5 | 10 oz basket | pint size berry baskets | baby zucchini and patty pan |
| Sunflower shoots | microgreens | $3/two for $5 | 2 oz bag | 5" × 8" bag w/twist tie | |
| Sunflower shoots | microgreens | $5 | 6 oz bag | 4" × 2" × 10" bag w/twist tie | |
| Swiss Chard | greens | $3/two for $5 | 8 oz bag | bunched with elastic band | |
| Tomatoes (cherry) | fruits | $3/two for $5 | 10 oz basket | pint size berry baskets | |
| Tomatoes (large heirloom) | fruits | $2.50/lb. | - | sold loose | |
| Tomatoes (Roma/ San Marzano) | fruits | $3/two for $5 | 13 oz basket | pint size berry baskets | |

**FIGURE 45.** Packing Chart for Chefs

| Product | Type | Restaurant Price | Unit Weight | Price per Pound/ Unit | Packaging | Item Details |
|---|---|---|---|---|---|---|
| Arugula | salad greens | $20 | 2 lb. case | $10 | 12" × 19.5" roll bag | |
| Arugula (box) | salad greens | $90 | 10 lb. case | $9 | waxed box with plastic liner | |
| Basil | herbs | $20 | 2 lb. case | $10 | 12" × 19.5" roll bag | |
| Beets (premium golf ball size) | root veg | $4 | sold loose by the pound | $4 | packed loose into waxed cardboard boxes | 2 inches of tops left and sold loose in boxes |
| Beets (larger size) | root veg | $2.50 | sold loose by the pound | $2.50 | packed loose into waxed cardboard boxes | all tops removed |
| Bok Choy | greens | $25 | 5 lb. box | $5 | packed loose into waxed cardboard boxes | waxed cardboard box |
| Braising mix | salad greens | $20 | 2 lb. case | $10 | 12" × 19.5" roll bag | |
| Carrots | root veg | $4 | sold loose by the pound | $4 | packed loose into waxed cardboard boxes | topped with 2–3" of green left |
| Cilantro | herbs | $2 | 2 oz bunch | $2 | bunched with elastic band | bunched in the field |
| Dill | herbs | $2 | 2 oz bunch | $2 | bunched with elastic band | bunched in the field |
| Eggplant (baby varieties) | fruits | $4 | / lb. | $4 | packed into tomato boxes | |
| Kale | greens | $25 | 5 lb. box | $5 | packed loose into waxed cardboard boxes | |
| Parsley | herbs | $2 | 2 oz bunch | $2 | bunched with elastic band | |
| Pea shoots | microgreens | $30 | 2 lb. case | $15 | 12" × 19.5" roll bag | |
| Radishes (6 lb case lot) | root veg | $25.00 | 6 lb. case | $5.00 | packed loose in a 12" × 19.5" roll bag | tops on or removed with 2 inches of green remaining |
| Radish shoots | microgreens | $20 | sold by the pound | $20 | 12" × 19.5" roll bag | |
| Red Russian kale | salad greens | $20 | 2 lb. case | $10 | 12" × 19.5" roll bag | |
| Salad mix | salad greens | $20 | 2 lb. case | $10 | 12" × 19.5" roll bag | |
| Salad mix (box) | salad greens | $85 | 10 lb. case | $9 | waxed box with plastic liner | |
| Salad turnips | root veg | $40 | 6 lb. case | $7 | packed loose in a 12" × 19.5" roll bag | |
| Spicy salad mix | salad greens | $20 | 2 lb. case | $10 | 12" × 19.5" roll bag | |
| Spinach (baby) | salad greens | $16 | 2 lb. case | $8 | 12" × 19.5" roll bag | |
| Spinach (large leaf) | greens | $12 | 2 lb. case | $6 | 12" × 19.5" roll bag | |
| Spring onions | root veg | $8 | sold loose by the pound | $8 | 12" × 19.5" roll bag for 1 lb or more | |
| Summer squash | fruits | $40 | 10 lb. case | $4 | packed into tomato boxes | |
| Sunflower shoots | microgreens | $30 | 2 lb. case | $15 | 12" × 19.5" roll bag | |
| Swiss chard | greens | $25 | 5 lb. box | $5 | packed loose into waxed cardboard boxes | |
| Tomatoes (cherry) | fruits | $40 | 10 lb. case | $4 | packed into tomato boxes | |
| Tomatoes (large heirloom) | fruits | $25 | 10 lb. case | $2.50 | packed into tomato boxes | |
| Tomatoes (Roma/San Marzano) | fruits | $30 | 10 lb. case | $3 | packed into tomato boxes | |

Packing items for restaurants is similar to packing for market. We work from a list of items to be packed in alphabetical order and portion each item as we go down the list. For example, starting with arugula, we will pack all of the case lot bags and boxes, then put them in the cooler. I sell items to chefs in standard case lot amounts because it makes packing simpler. Once we have gone through the portioning list, we then pack the specific orders. I will lay out boxes in an open and shady area, label each of them with painter's tape, then go down the list and allocate all of the items into each designated box. Once all of the orders have been put together, I then load the truck or bike trailer. Then I print invoices; I wait to print invoices just in case there are slight changes (like being short on a certain item) to the orders.

When we are packing restaurant orders, we try to be exact with the weight—or slightly over—never under. Since these customers are buying large volumes, we want to deliver the perfect amount.

PART 8

# PRODUCTION SYSTEMS

In this part, I will outline all of the tasks related to production on my farm. Some of these tasks are common for many small farms, and some are totally unique to the urban farm. Because of the size of my farm, some production methods are superimposed because I usually have to deal with lack of space, so some of my techniques can be a little unorthodox.

# Beds for Production

On an urban farm, our bed unit is a little shorter than most small-scale intensive farmers because of the small plots we're working on. If I had the space for 100-foot beds, I'd happily use them because I'm often planting and harvesting that amount of product on a weekly basis. Unfortunately, it's rare to find an urban plot with dimensions that would hold a 100-foot bed. However, I prefer to use a shorter-sized bed, and there are many reasons for this.

Thirty-inch width has become standard for small-scale intensive farmers, and it is often the width of small machine implements. One very important benefit of a bed this size is that it is narrow enough to straddle. This is very helpful when you are working the beds so often—whether it's standing over them to harvest, or kneeling down to weed or do any other tasks. A narrow bed can be reached from both sides, and there are many different positions one could use for working such beds. I find this helps to create an ergonomic workflow. I like to be able to work a bed from different positions, so my body doesn't get sore from being in the same position all the time. See photos #31–33 in the photo insert section; they show different ways to work a bed.

Beds that are four feet wide (often what small farmers use), are hard to stand over, let alone walk across. If you're managing a farm with hand tools, a four-foot-wide bed is a sure way to throw out your back. Some farmers refer to four-foot beds as *double reach beds,* because you can reach in from both sides. The problem here is ergonomics. Do you actually want to hold your body in an outstretched position day in and day out? I know I certainly don't. I have a horrible back after nine years of treeplanting, and I want to make everything on my farm as easy on my back as possible. For me double reach is double back pain.

Regarding the length of a bed, using numbers divisible to 100 is the best approach for a variety of reasons. First of all, most irrigation and season-extension equipment comes in lengths divisible by 100, so having your beds be in 25s, 50s or 100s is ideal. One 25- or 50-foot bed also typically contains a volume of produce that I can move on a weekly basis. Firstly, it's better to crop out entire beds at a time so that a crop (like in the case of Cut and Come Again Greens) is consistent in its length if it grows back. Secondly, you can crop out the entire contents and immediately plant something else in its place. Efficient use of every bed is not as much of an issue for anyone farming over an acre, but for me (on ⅓ acre), all of my beds need to be in production all the time.

For example, if I had a 100-foot bed of a crop and I harvested 25 feet of it each week, it would be four weeks until I could turn that bed over to plant something else. One 25-foot bed of radish is 70 bunches; I can easily sell that in one week. A fresh cut of lettuce greens (20 pounds) or a bed of scarlet frills mustard greens to be part of a salad mix is the same.

## Units of Production

Perhaps the most useful application for standardizing bed size is to use beds as a units of production. We create standards and benchmark figures through our Hi-Rotation and Bi-Rotation beds to easily calculate how much production each piece of land could generate. For the urban farmer, it isn't about how much land (in acres or even square feet) you have to farm, it's about how many beds can you fit into each space.

## Unique Beds

There's always a reason to break the rules, and here is a variety of reasons we would break them.

### *Odd-Shaped Plots*

A lot of the time, the plots I'm farming aren't an ideal size. It's rare to find a piece of land that is a perfect square (50 by 50 feet). In most cases, an urban lot is a perfect square or rectangle, but there are physical obstructions that prevent you from using ideal dimensions. Trees, balconies or sheds are often in the way, so there are going to be times where you just can't create beds in increments of 25. For example, in my front yard, the plot is 22 by 35 feet, and it's an oblong shape.

It's better to have fewer, longer beds, than a series of short beds, because that means there will be less space dedicated to walkways and more for beds. In some cases, short beds can be logical. See photo #34 of my front yard in the photo insert section.

### *Short Beds*

Sometimes using short beds can be a great way to grow crops that you don't use a lot of; for example, it could be part of a salad mix where you need only a few pounds or less on a weekly basis. I have used short beds for crops like micro carrots (something I grow small amounts for certain chefs) and for field microgreens. Short beds can be a great way to make good use of an awkward area. Even when I make short beds, I try to shape them in 25-foot divisions (like ½ beds (12½ feet) and ¼ beds (six ¼ feet)). Quarter beds are what I use for all of my field based microgreens, and I use ½ beds for salad mix ingredients and one-off crops. See photo #35 in the photo insert section.

### *Double Beds*

There are times when doing away with the walkway between two beds of the same crop can give you extra rows of production; for the urban farmer, every inch counts! In my experience, double beds only work for Quick Crops in HR areas, where you plant a particular crop in multiples of two and they will be out of the ground fast. For example, if you're planting two beds of radishes each week, cropping them out—and they're being planted next to each other—you can till or fork up the walkway between the beds and plant extra rows in there. The walkway isn't needed in this case because you're going to harvest the crops all at once, and you're not going to need to work the beds after planting (in most cases, radishes grow faster than weeds).

You could plant any Quick Crops that are harvested at once in double beds, but I wouldn't recommend this for Steady Crops or even Quick Crops that are harvested more than once (like cut lettuce greens). The key is that a double bed is best for one-off quick-growing crops. See photo #36 in the photo insert section.

In the examples below, you can see how by making a double bed out of two beds at a 7 rows per bed, you can get another three rows in between. That's an extra 21% production. In the case of two nine-row beds, you can add another four rows. That's an extra 22% production.

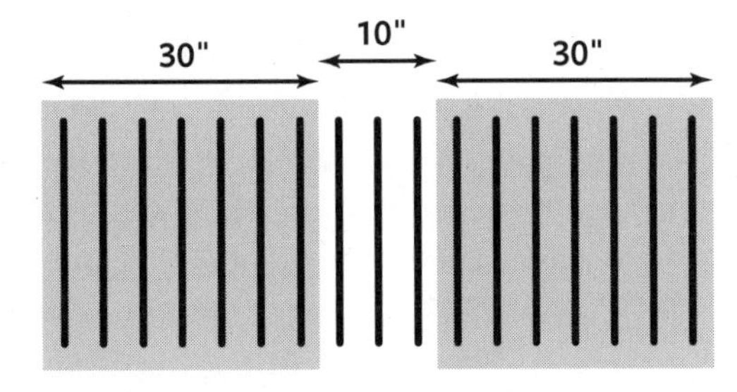


**FIGURE 46.** Two beds of radishes at seven rows each, side by side. If you use the walkway to create a double bed, you can fit another three rows in the walkway space. Credit: Curtis.

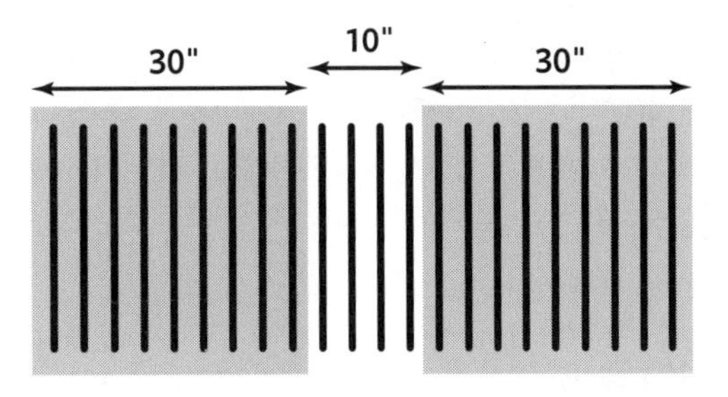


**FIGURE 47.** Two beds of arugula at nine rows each, side by side. If you use the walkway to create a double bed, you can fit another four rows in the walkway space. Credit: Curtis.

### *Long Beds*

I consider anything 50 feet or more to be a long bed. I have some plots where I have all long beds, just because having twelve 50-foot beds makes better use of the land than 24 25-foot beds. In some cases, I'll approach a long bed as I do a double bed: if I'm planting enough of a certain crop each week to justify a longer bed, sometimes I'll connect two separate segments and make a temporary group of long beds, much the same way as I approach the double bed. See photos #7, 10, 34 and 52 in the photo insert section.

### *Interplanted Beds*

Another way to maximize the production on an urban farm is to look for opportunities to double up crops that won't interfere with each other in the same area. Some organic growers approach interplanting on a needs-and-benefits basis for the crops: pairing a nitrogen fixer with a nitrogen dependant, or a plant that repels the pests that one plant attracts. Though I do think there is merit to this approach, it's not why I use interplanting. I pair short season with long season in order to get overlapping production out of the same area. Two examples of common interplanting strategies I've used on my farm follow below.

#### Tomatoes

When tomatoes go into the ground in my greenhouses, they are interplanted amongst early season crops such as lettuce, arugula, radishes and spinach. With some beds that are harvested at once, like radishes, we would use the whole bed to plant tomatoes while interplanting basil amongst the tomatoes as well. With other crops which are cut multiple times (like arugula, spinach or Red Russian kale), we sometimes forfeit a row or two of the crop to make room for a row of tomatoes. Once you do this, you must be very careful when harvesting your greens to make sure you don't cut your tomatoes. Once the tomatoes are about four feet tall, all other spring crops will be removed as the tomatoes will require more room to grow properly.

When you are interplanting into a small greenhouse, you must step very carefully. In some cases, we have interplanted tomatoes in the walkways between crops such as radishes and turnips. Where you plant depends on where the support wires for your trellising are set. In my 18-foot-wide wide tunnels, we put in four rows of tomatoes: two on the outside, about three feet six inches from the sides, with two rows down the center 18 inches apart. Then we further interplant basil with the tomato rows that have room for them. In the 12-foot-wide tunnels, we put in three rows of tomatoes that are right in the walkways between the four beds of spring crops. On the sides, the rows will be closer to the edge of the beds, so that there is some room to walk as you're harvesting the remaining spring crops. This type of intensive planting is definitely not for everyone, as it can be a little tricky to manage for the two- to three-week period where the spring crops are crossing over with the tomatoes. See photos #21, 22, 24, 37 and 43 in the photo insert section.

**Pattypans**

A great way to utilize the space between pattypan squash plants in a BR area is to plant between the rows of pattypans. I plant pattypans in 30-inch beds with one-foot walkways in between; the rows are in the center of the bed, so there is 42 inches between each row of pattypans. When these squashes first get transplanted into the ground, you can rototill a 30-inch strip down the middle, with six-inch walkways on both sides, and plant a very quick-growing crop like radishes, spinach or arugula there. Those interplanted crops will be ready to harvest in 21 days, which is just about when the squash plants will be big enough to need more space. This doubles up a Bi-Rotation area with an extra crop!

**FIGURE 48.** Here's an example of interplanting a Quick Crop with a Steady Crop. The small plants in the center of the drawing will mature and be harvested before the pattypan squash is large enough to interfere.

## Turning Over by Tillage

To till or not to till, that is the question. Today, in the organic and small farm community, tillage has gotten a really bad name, and for some good reasons. There are many adverse effects to tillage: damaging soil microbes and worms, creating hardpan and inverting soil layers that bring up weed seeds. I have seen many of these effects myself but not as extensively as they are made out to be.

The reality for a small farmer like me is that using a rototiller is just a practical way to get the job done a lot faster. If it weren't for my tiller, there would be a lot less chance that a farmer like me would be able to accomplish as much as I do on a daily basis.

Tilling has been beneficial for my operation because I can turn over new plots of land and change the physical composition of poor soils quickly. I don't think that tilling should be a long-term strategy by any means; however, some urban farmers are frequently not on any piece of land for more than three years. Our garden plots may either go back to lawn or be developed into homes. I make all that I can happen on someone else's land in a short period of time. If that's the case, then running a rototiller to turn over your beds on a weekly basis will likely not have significant negative long-term impact.

The primary reason I no longer till for consistent bed rotation is because of weeds. Each time you invert the soil through tillage, you bring up new weed seeds from underneath. The top layer of soil is often peppered with weed seeds, and each time you turn it, you bring up new ones. If you're having to manage weeds on a weekly basis, this can create a lot more work for you. With some Quick Crops like radishes and turnips, weeds aren't as much of an issue because these crops grow faster than weeds; they're most often harvested and out of the ground before those weeds can reproduce themselves. However, with Quick Crops of greens that will be cut multiple times, weeds hinder production. Having a lot of weeds in beds of arugula or lettuce means that you will be spending time sorting weeds from crop after your harvest.

### *How to Turn Over Beds With a Tiller*

When turning over your beds with a tiller, it's best to have a machine that is the width of your beds. First, pull or fork out any deep-rooted weeds before you till, otherwise they will just come back. Make sure there isn't too much crop residue from the previous crop, because it can create problems for you next crop; residue can also make tilling more difficult as some of it will get tangled up in your tiller tines. In some cases it's better to pull the previous crop and compost it than it is to turn it under. Don't till in full lettuce beds that are five inches tall if they have gone to seed, because you'll have to till the beds so many times to turn it all under that it may be less effort to pull the plants before you till. However, small or modest amounts of crop residue are perfectly fine to turn under. Once you've established whether you're going to remove the previous crop or turn it in, then it's time to add your soil amendments, then rototill those in.

Tilling on urban plots can sometimes be tricky, navigating as you are often between physical obstructions such as walls, hedges or fences. You need to be careful to not back yourself into corners; always give yourself enough room to turn around. In situations where beds run up into a fence, I'll stop the tiller three feet before I reach the end of the bed, and turn around 180° before the end, then drop the tiller back down and proceed. This does mean that that one end of the bed may get tilled a little less than the rest, but it barely makes a difference.

When I turn over beds with the tiller, I'm turning over only the beds themselves; the walkways and perimeters remain static and are not tilled. After a couple of years of tilling, using a broadfork or a pitchfork to loosen up the subsoil in your beds will help avoid subsoil compaction. See tilling photo #38 in the photo insert section.

## No-Till Beds

A lot of new farmers are starting to develop strategies to minimize tillage or eliminate it all together. The challenge up until quite recently has been how to implement these ideas on a small farm. No-till methods have been around for many years on some large-scale farms. They have been using large tractor implements that can knock down the previous crop and drill seed right into the ground afterwards. For small farmers, tilling creates clean seedbeds that a small seeding tool can carefully and evenly drop seeds into. There are still no tools that I have seen that can direct seed into a bed with mulch. If a clean seedbed is the goal to run a seeder, there are three ways I will describe in which this can be achieved without tillage.

With all of these methods, if you're preparing your beds for the first time in the season, you should use a broadfork or pitchfork to first loosen the subsoil on your beds.

### *No-Till Using Non-Mechanized Hand Tools*

This is a technique I have been using for years. It works and it's cheap, but it's hard labor, and in the summer heat can be a real drag. However, on a small urban farm of ¼ acre or less, hand tilling can be done economically.

#### Step #1: Remove the Previous Crop Residue

Pull out all the previous crop from the bed. You must remove it all, because hand tilling will not work if there's too much residue. Pull from the base of the plant and shake the soil off so that you're not moving soil around. It's best to keep as much soil intact as you can.

#### Step #2: Stirrup Hoe the Roots and Loosen the Top Inch for a Fluffy Seedbed

Take a stirrup hoe and run it up and down the bed, loosening up soil along the entire bed. This is essentially like rototilling by hand, at a very shallow depth. This way we can achieve a fluffy topsoil that we can later level out and seed again. Make sure you're hoeing every square inch of your bed's surface.

#### Step #3: Sprinkle in Soil Amendments and Rake out Leftover Debris

Next, add appropriate soil amendments for the bed and rake them in. If you walk with your rake on an angle, with both of your hands in a "thumbs up" position, you will be raking the bed level, mixing in your amendments, as well as raking any leftover debris to the edge of the bed. Debris will end up in the walkways, and that's totally fine.

Now your bed is ready to plant again. Once you get good at this technique by doing it many times, you should be able to turn over a 25-foot bed in 30 minutes.

### *No-Till Using the Tilther*

For this technique, you need to very specialized tool called a tilther. It is an electric drill-powered mini tiller that is 18 inches wide. The tilther doesn't rototill in the traditional way in that its tilth is very shallow—only about an inch. The main purpose of this tool is to prepare a new seedbed in a similar way to a tiller, but a tilther fluffs up the top layer of soil, so that your seeding tool can operate smoothly through the bed and so you can work soil amendments into the surface of the soil at the same time. In order to use this tool effectively, your bed must be free of crop residue.

1. Remove the previous crop residue. Same as in the previous method.
2. Sprinkle on soil amendments
3. Use the tilther to mix in amendments. This may take a couple of passes over the bed, depending on soil conditions.
4. Make one pass with a landscape rake to level out the bed

See photos #39 and 40 in the photo insert section.

#### *No-Till with a Power Harrow*

The power harrow is a walk behind tractor implement similar to a tiller, except that it doesn't invert the soil. It has tines that spin on a horizontal axis; it scuffs up the soil only near the surface inch. This helps prepare a clean and level seedbed. When using a power harrow, you (1) clean up previous crop residue, (2) add your amendments, (3) then use the machine to prep and level the beds. No raking is required. This tool can be difficult for use on urban farms because it is very large and heavy; it is very difficult to maneuver in tight spaces.

# Planting

I plant in three different ways on my farm. The first, and most common, is *direct seeding.* I direct seed most of my quick-growing crops almost all of the season. The exceptions to this are early crops like bok choy and first planting of beets, head lettuce and scallions. Once it's warm enough, those crops can all be direct seeded, except head lettuce. I mostly plant that in plugs.

The second planting method is *transplanting.* All of my Steady Crops such as tomatoes, pattypans, peppers, chard and kale are started in the nursery early in the season, then planted out to the field as it warms up outside.

The third method is *hand planting.* Hand planting is direct seeding by hand at a very specific density. On some occasions I will plant fall kale, beets and head lettuce by hand.

## Direct Seeding

For most of the growing season, direct seeding happens nearly every week on the farm. The crops I seed most often are my main Quick Crops such as arugula, spinach, asian greens, radishes and turnips. Carrots and beets are also direct seeded but in slightly wider intervals than the others, mainly because they have a much longer date to maturity but also have a longer window to harvest.

My most common seeding tool is the Jang seeder I described in Chapter 27.

### *How to Plant Perfect Rows By Eye*

Some farmers prefer to mark out rows in a bed before planting. I, however, plant everything by eye. If you are planting a bed with an odd amount of rows (three, five, seven or nine), this is very simple. Always start with the outside rows. Start outside the bed; you don't want to plant right to the edge of the bed, only about two inches in from each edge. Plant the outside rows on each side of the bed, then, to plant three rows total, split the difference right down the middle. To plant five rows, first plant three, then

**FIGURE 49.** Recommendations For Seed Sowing Density: DTM=Date to Maturity, GH=Greenhouse, PLT=Poly Low Tunnel

| Crop | Avg DTM from seed date | When to Seed | First GH/ PLT Plant Date | First Field Plant Date (no cover) | Last Field Plant Date | Last GH/ PLT Plant Date | Jang Seeder Roller | Earthway Seeder Plate | Seed Depth | Rows/ Bed | Seed per 25' bed Grm | Seed per 25' bed Oz |
|---|---|---|---|---|---|---|---|---|---|---|---|---|
| Arugula | 35 | Mar–Oct | Mar 9 | May 4 | Sep 8 | Sep 22 | YYJ24 | — | ½" | 9 | 11.6 | 0.41 |
| Beets | 72 | June–Aug | — | May 11 | Aug 3 | — | — | chard | ½" | 4 | 5 | 0.18 |
| Beet greens | 40 | May–Aug | — | Jun 1 | Aug 3 | — | — | chard | ½" | 7 | 83 | 2.93 |
| Carrots (baby) | 65 | Apr–Aug | Apr 13 | May 4 | Aug 3 | Aug 10 | XY24 | — | ½" | 7 | 7 | 0.25 |
| Carrots (full size) | 78 | Apr–Aug | — | May 4 | Aug 3 | Aug 10 | XY24 | — | ½" | 5 | 5 | 0.18 |
| Cilantro | 30 | Apr–Sep | Apr 6 | Apr 13 | Aug 17 | — | G12 | chard | ½" | 9 | 78 | 2.75 |
| Dill | 60 | Apr–Aug | Apr 6 | Apr 13 | Aug 3 | — | MJ24 | spinach | ½" | 9 | 24 | 0.85 |
| Salad turnips | 38 | Apr–Sep | Mar 16 | Apr 27 | Aug 17 | — | YYJ24 | — | ½" | 9 | 5 | 0.18 |
| Lettuce | 45 | Mar–Oct | Mar 16 | May 4 | Aug 31 | Oct 5 | F24 | — | ½" | 9 | 28 | 0.99 |
| Mustard | 35 | Mar–Sep | Mar 9 | May 4 | Sep 8 | Sep 15 | X24 | — | ½" | 9 | 22 | 0.78 |
| Parsley | 70 | June–Aug | | May 11 | Aug 3 | — | MJ24 | spinach | ½" | 5 | 15 | 0.53 |
| Radishes | 28 | Mar–Sep | Mar 16 | May 4 | Aug 25 | — | F24 | — | ½" | 7 | 20 | 0.71 |
| Red Russian kale | 30 | Apr–Sep | | Apr 27 | Aug 10 | Aug 31 | F24 | — | ½" | 9 | 17 | 0.60 |
| Spinach | 45 | Mar–Oct | Mar 9 | Apr 6 | Sep 8 | Oct 5 | — | chard | ½" | 5 | 54 | 1.90 |
| Tatsoi | 35 | Mar–May, Sep–Oct | Mar 9 | Apr 13 | Sep 7 | Oct 5 | X24 | — | ½" | 9 | 17 | 0.60 |
| Scallions | 70 | May–Aug | — | May 4 | Aug 3 | — | F24 | — | ½" | 7 | 15 | 0.53 |

split the difference between the three to make five. To plant seven rows, plant three and then plant two rows between the three rows. Just make them equal distance apart. Planting nine rows works the same way as planting five, but split the difference twice: plant four rows in between each of the five rows you just planted. With all of these odd number combinations, always start with the two outside rows.

Most of my crops are direct seeded at uneven rows. There is no particular reason for this, except that those densities have worked great for me—and planting odd numbers is simple. However, you may find a crop at some point that you want to plant at an even number of rows, and this is a little less simple, but not complicated.

For example, to plant four rows in a bed, simply, plant the outside rows of the bed first, then plant the next two rows by splitting the difference down the middle. Sometimes, I use my hand to mark a line down the middle, then slightly offset rows from that middle line. If your rows are not exact, it's not a big deal. Planting is not rocket science! See photo #41 in the photo insert section.

### Transplanting

Transplanting is the planting of seedlings that were started in a nursery. The main reason for transplanting is to get a head start on a crop, so that the plants have less time in the ground. This is necessary when you need to turn over areas quickly, and it is critical for an urban farm on a small land base. I have no areas on my farm that are single rotation (with just one crop in them for the season). There are always at least two rotations of every bed. With the use of transplants, I can have an early Quick Crop in the ground and out before another slower-growing transplant (such as tomatoes) goes in. It can also be the other way around. In the case of kale, early transplants are set out; then that crop is finished by midsummer and something else goes in its place.

Before we transplant into the ground, our beds are either prepared using the stale seedbed technique for pre-emergent weeding, or they are covered with a landscape fabric with holes punched in where the transplants will be placed. When the beds are planted without landscape fabric, we use a landscape rake with small plastic pipes attached to the tines, to mark rows—or we use a tool called a seedbed roller. See transplanting photos #42, 43 and 44 in the photo insert section.

### Nursery

The nursery on my farm operates on a micro level compared to most farms. Because we focus primarily on quick-growing crops and most of our planting is done through direct seeding, our nursery infrastructure is minimal. See Chapter 28 about nursery infrastructure, and photos #13–16, 45 and 46 in the photo insert section.

#### *Soil Blocks or Plugs*

A lot of farmers have been moving towards *soil blocks* for starting, and having used them myself for a couple of years I can see why. They are superior in many ways; however, they are far more time-consuming to make. Most farmers use standard nursery flats of various sizes (with 200, 128 or 72 cells per flat). They're easy to fill with soil and are readily available. The problem with flats is that your crops can become root-bound if they are in the cells too long. (This means that the roots will start to spiral around in the cell, and when it comes time to plant, the plants will take a lot longer to continue growing as they direct their roots into the ground.) Soil blocks address this issue because the roots air prune themselves as they grow to the edge of the block. A soil block is essentially a stand-alone block of soil without any cell or pot holding it in place. Soil blocks hold their form because of the way the soil is mixed together: adding the right amount of water to make an almost mud-like mixture that is packed into a cookie-cutter-style mechanism that makes small blocks of soil that are separated by air. When the roots grow to the edge of the block, they split off like capillaries in your lungs, which actually creates more surface area of root matter, and that helps the plants grow when they are transplanted into the ground.

I prefer a bit of both. I use plugs for all of my long-season crops like tomatoes and summer crops, but I prefer soil blocks for ongoing nursery stock like lettuce, bok choy, beets and even basil. For these, I use a mini blocker; I like it because I can fit 420 blocks in a 10 by 20-inch flat. When it comes to transplanting plants in soil blocks, it's actually faster because less time is needed to pop them out of the flats. For a list of nursery supplies and my soil mixture for plugs and soil blocks, refer to Chapter 28. See photo #46 in the photo insert section.

# Microgreens

Microgreens have been growing in popularity over the past few years, especially as more farmers are seeing their profit potential. It's true that micros are very high value and an incredible amount of production can come off a very small space, but as any investor would tell you, the faster and higher the return, the higher the risk. Seed is a big expense when it comes to microgreens, and if you are not selling all that you are producing, these greens can become a money-losing endeavor very quickly.

To introduce them into your crop portfolio and customer base, start small and test the waters. Only scale up when you see the demand growing. If you are composting a lot of micros on a steady basis, you are most likely losing money. Microgreens are still a pretty new phenomenon for the average farmers market customer. When bringing them to the farmers market, you'll have to do a considerable amount of educating to get your customers to learn how to use them and to appreciate their health benefits as well.

I grow microgreens in two ways on my farm:

1. In one-inch-deep germination flats indoors or in greenhouses
2. Directly in the soil with a technique I have developed called *The Board Technique.*

The three main microgreen crops I grow are pea shoots, sunflower shoots and radish shoots.

## Indoor and Greenhouse Growing

Indoors I grow microgreens on a multi-level shelving unit equipped with fluorescent grow lights. With this method, you must take caution to control your humidity levels and heat and make sure you have constant airflow; otherwise you will encounter fungus issues that can cost you entire harvests. We grow indoor from November to early April.

After these months, as nighttime temperatures warm to above freezing, I can move the flats out to my high tunnels, where they will grow on vertical T-hangers from the ridgepoles of the tunnels. When the flats are planted, they will still be placed indoors on the shelving unit to germinate and emerge; this takes between five and seven days. See photos #14 and 47 in the photo insert section.

There are nine stages to growing microgreens in flats:

1. Sterilize the seed
2. Soak for six to eight hours
3. Drain and rinse
4. Prepare soil for planting
5. Fill the flats and water
6. Plant
7. Cover with sterilized empty flats, then stack to germinate
8. Uncover after emergence and expose to light
9. Harvest and wash

1. Sterilizing seed is most important for sunflower seeds, as they have the greatest chance of fungus developing. Before I soak the seed, I will mix four teaspoons of white vinegar and four teaspoons of food grade hydrogen peroxide in one quart of water and soak the seeds for ten minutes. Then drain and rinse them. Some health authorities advise that all soaked seed be sterilized. For any seed that is dry when planted, like radishes, you can skip this stage.

2. After sterilizing our seeds, we soak them for between six and eight hours. I usually do this before I go to bed, then rinse them in the morning. For pea shoots, I soak 12 ounces of seed per flat, for sunflowers six ounces per flat.

3. Thoroughly drain and rinse the seeds after they have been soaked, then put them in a clean bucket. If you can't plant them right away, you can delay this process by a day or two by continuing to rinse and drain them a couple of times a day and keeping them in a shady and cool area.

4. To prepare soil for planting. I use my soil sifter made from ¼-inch steel mesh held in a small wood frame that is built to fit on a tote. I shovel new sterile soil mix on the sifter, and push it through with my hands. This removes any large chunks of debris to make a very fine soil for planting. I will usually sift many totes at a time, so that when it comes time to plant, I can skip this stage.

5. Now I fill each one-inch-deep flat with three quarts of finely sifted soil. (For sunflowers, I sometimes will use two-inch deep flats, especially when growing indoors. I have found that they yield slightly more and are less prone to fungus with a little more soil. I use 3.5 quarts in this case) Then I level out the soil on the flats with my hands. Now, I take a small piece of plywood

with a handle on the top, that is the dimension of the flat, and I press down the soil to make a firm planting surface. Then, using a fine watering wand, I water the flats heavily. The best way to tell if you have watered enough is to stick your finger into the soil on a corner of the flat, and if it's wet all the way down, you've watered enough.

6. Next I plant the flats. Depending on how much seed you've soaked, spread the seed evenly amongst your flats, then spread it around with your hands. In the case of radishes, sprinkle the seed on dry. I use five tablespoons of radish seed per flat. See photo #47 in the photo insert section.

7. Now cover the seed with empty flats and stack them to germinate. It's important to use flats that have been properly cleaned and sterilized when doing this. Contaminated flats may cause fungus problems during germination. I place the flats on my shelving unit with one layer of planted microgreens, an empty flat directly on top of the seed, then a sheet of corrugated plastic on top of that. Then you can place other layers of flats on top of that in the same way. I never stack more than three layers. On the top layer, I place some books for weight.

8. Once the crops have emerged, lift up the flats that were covering them and bring the sprouted seeds into the light. I wait one day or a half day to water them, as there is still usually a lot of moisture in the soil. During warmer months, if I'm still growing in flats, I will uncover them and keep them out of direct light for a day or two to allow them to get elongated, then I will put them in the direct light.

9. After between five to seven days of being uncovered (depending on the season), the microgreens will be ready to harvest and wash. For details on harvesting, see Part 10. For how to wash them, see Chapter 33, Post-Harvest Processing.

### Field Micros and the Board Technique

Over the years, I have experimented with faster ways of growing microgreens. I often found that, once more field crops came into production, finding the time to plant flats of microgreens became more difficult; but the demand stayed, so we were fighting to keep them going. At first, I started to direct seed them in the ground, using the Earthway seeder; this worked to some degree, but the dates to maturity (DTM) were inconsistent during the spring and fall—plus, the yield was very small per square foot of area, compared to growing in flats. This was mainly because germination times during colder nights were so variable. Since I was already soaking and germinating my seeds indoors for flat production and growing them in my high tunnels on the T-hangers, I figured why not combine the two techniques and use some of the aspects of growing in trays to growing on the ground?

The main benefit of growing field micros is that they take considerably less time to plant. One six-foot bed (which is the equivalent size of ten flats) can be planted in minutes. Also, we no longer need to buy soil mix because we are using the soil on the ground. Another great benefit is that all this crop residue helps build massive amounts of organic matter into our soil. Microgreens do not pull nutrients from the soil. A plant at this stage is solely relying on stored energy from its seed and the first photosynthesis that occurs. Because of this, I have been able to plant in the same beds over and over by rototilling or hand turning in leftover residue and replanting. Over time all of that residual organic matter has helped to build up the structure of my farm's soil. So, in a way, I am getting the benefits of a green manure while making a good healthy profit at the same time. I have planted and rotated field micros in areas with poor soil before, as a proactive way of building up the organic matter, and it has worked incredibly well. However, I would not advise using this method on any soil that may be contaminated.

The board technique uses 30-inch by six-foot sheets of plywood with two six-foot two-by-fours on the sides of each board (to make it easy to pick up and keep the sheet straight without bowing over time). We plant on a 15-square-foot area (roughly the equivalent of ten flats of micros in 10-by-20-inch flats). The seed is soaked and rinsed exactly the same way, and the board is placed over the seeded bed in the same way as empty flats are placed on top of newly planted micros. The boards essentially help germinate the crop quickly and establish a perfectly level crop. I place a board on top of the newly seeded crop and lift it up after four days. Then, the board is taken off, the crop is exposed to light and will be mature in between five and seven days, depending on the time of year.

The stages of growing field micros are nearly the same as flat-based micros, and I will highlight only the stages that are different:

1. Sterilize the seed
2. Soak for six to eight hours
3. Drain and rinse
4. *Prepare soil for planting*
5. *Tamp down and water*
6. *Plant and sprinkle a light layer of soil*
7. *Cover with the board, tamp down and water again.*
8. Uncover after emergence and expose to light
9. Harvest and wash

The only stages that are different about this method are #4, 5, 6 and 7. All other stages are the same. Harvesting is also done in the same way, except that you're cutting from the ground.

4. To prepare the soil for planting, rototill or turn it by hand. There's nothing different here from usual bed prep techniques, except that we don't have to worry about amending the soil. If, however, you are going to replant where field micros were previously, you need to make sure your seedbed is free of residue on the surface. This can be

done through no-till, by stirrup hoeing the last of the previous crop, turning it under, then replanting from there. Personally, I prefer the tiller because it is much faster and we are not concerned about weeds here at all; the microgreens grow far faster than any weeds will. I will often prepare entire 25- to 50-foot beds for field micros at a time, so I'll be tilling at least four six-foot beds at once.

5. After the bed is prepped, take your board(s) and lay them on the soil. Walk on them a bit to press down the soil surface to get a firm and flat bed. Just like with flats. Then, just as with flats, heavily water the soil with your watering wand.

6. Now, spread around your seed just as you do with flats. The only difference is that in the field you'll dump down a lot more seed—otherwise the process is the same. After that's done, spread a small bucket full of finely sifted soil over the seeds; you can use soil from the ground or your peat-based soil mix. I do this because I have found that I get a better germination; the soil helps fill gaps between the seed to mimic what would happen naturally. Indoors, this doesn't need to be done because temperatures and humidity are much more stable.

7. Now cover the bed back up with the board, walk on it again, then lift the board back up to water the bed in again. Now, cover it up with the board—and leave it for between five and seven days.

See photos #48 and 49 in the photo insert section.

# Extending the Season

Extending the growing season is critical for maximizing production on a small land base. In this part I'll describe techniques and equipment you can use to push your season to the edges of winter, and even over the winter.

When it comes to season extension, farming in the city has some inherent advantages. The *heat island effect* gives farmers in cities up to 22°F warmer nighttime temperatures than farmers right outside of a city. This means that urban farmers can often produce more at the beginning of the season than others. Even though my farm is very small compared to most, I have more for sale at the first few farmers markets than most other growers, and that's even farmers that are on acres of land.

The first and most important thing to understand is that certain crops do better at certain times of the year. For example, putting tomatoes out in the ground early, even if under poly, isn't always the best strategy for extending their growing season. Extending doesn't necessarily mean just having tomatoes for longer or earlier than everyone else, though that's part of it. At cool times in the year, it's better to focus on what grows best at that time of the year. Early in the season I focus on crops such as arugula, tatsoi, mustard, spinach, radishes and turnips. In many North American bioregions these crops will do very well during a cold spring and will even survive mild frosts.

## Poly Tunnel Greenhouses

Poly tunnels, when added to the benefit of the heat island effect, help bring a lot of early production off your fields. High tunnels have a greater volume of air inside, and the temperature fluctuations are less radical than in low tunnels. However, they are a lot more expensive, and you can't always put a structure like a greenhouse in someone else's backyard. See poly tunnel photos #50 and 51 in the photo insert section.

## Poly Low Tunnels

The best way I have found to extend seasons on my farm is to use a poly low tunnels. See photo #52 in the photo insert section.

Using these tunnels, I can have my entire ⅓-of-an-acre farm under poly at the beginning and near the end of the growing season. These tunnels allows me to have the benefit of greenhouses without being restricted by where I can put them. Low tunnels also allow season extension when you need it, but as summer comes on and you no longer need to cover your crops, you can take the tunnels off and leave crops in the open ground.

Low tunnels require quite a bit of labor to set up during spring, and some labor to maintain as you get snow in the fall. Your beds must be fully prepared and planted before you cover them up. Watering beds underneath low tunnels can also be a challenge: you must open them at least partially up to water them.

Using drip irrigation underneath the tunnels will save you a considerable amount of time at the beginning of the season. Poly low tunnels in my BR areas all have drip installed, and the low tunnels in HR areas are hand watered for the first couple of weeks because those areas are on overhead irrigation. Fortunately, when temperatures are cool not a lot of watering is required.

**FIGURE 50.** Hand Watering Poly Low Tunnels in the Spring.

However, on days where temperatures rise, it is best to open low tunnels up so they can ventilate properly. You must be careful with crops that have a tendency to bolt under stress (like bok choy and radishes). Sometimes I find it best to clip one end of each tunnel open so they don't overheat in situations like this.

How you space low poly tunnels is especially important in the fall. If you're using them for overwintered crops like spinach or carrots, it's best to leave one bed fallow between your tunnels. Once you start to get some snow, if your tunnels are too close together, that snow will build up between the tunnels and force the plastic in enough that it will smother your crop. If you leave a bed between, it is easier to manage snow as it piles up.

### Using Tunnels in Warm Climates

The term *season extension* can also be applied to warm climates such as southern California and Florida where the summers are often too hot to grow anything, or evaporation rates are so high that it's hard to keep moisture in the soil. All the same techniques described here can be applied in these climates by using shade cloth on top of greenhouses or poly low tunnels. I believe that drip irrigation is the best approach to watering in hot climates; therefore I'd recommend a combined strategy of shade cloth and low tunnels accompanied by drip irrigation to keep growing during hot summer months.

### Winter Farming in Cold Climates in North America

Farming in climate zones of 7 or less during the winter is more a figure of speech than it is a reality. Technically if you're producing something and selling it, you're farming; but for the most part, crops in cold climates will not grow during the winter, but they can be kept alive and harvested. This section is relevant for climate zones from 5–7.

The key is to have a strategy in place of what exactly you can have available during this time and how much you are going to harvest each week. Since the crops you harvest (like spinach, kale and sometimes lettuce) will take so long to regenerate after being cut, you need to know exactly how much you are going to harvest each week as if they were never going to regenerate. All you have to plan is one group of plantings. You will not be able to succession plant as you can during the main growing season. You may be able to stagger your plantings a little at the beginning, but you will not be rotating beds during the middle of the winter. There is just not enough daylight and heat.

Farming over the winter on my farm has not been a full-time endeavor. It's something I have done to keep some cash flow coming in, but it's very minor compared to that of the main growing season, mainly because you just can't get anywhere near the same level of production off the land. The nice thing about winter farming is that there is very little work to do on a weekly basis. Once the crops are established, there is very little consistent planting or maintaining that needs to happen; when the crops reach maturity, the cold sets in and then these crops pretty much just sit dormant. The main work outside is keeping snow off your tunnels and harvesting.

The equipment and infrastructure you need for winter farming is not much different than what you would need for farming in general, except that you must have an indoor area to process and pack your veggies, otherwise it's going to be very unpleasant washing or doing any other kind of wet work. You must have some poly high tunnels and/or poly low tunnels. High tunnels are much more enjoyable to work in than low tunnels as you are in an isolated environment and don't have to worry about moving snow around.

As for watering, if your winter temperatures are hovering around freezing, you won't need to water any of your crops because they should be mostly mature or at least well established before real winter sets in. The general rule is that if temperatures are 50°F or lower, you shouldn't water, as it may do more harm than good.

I approach winter farming by creating a production plan based on how many beds I'd like to have planted out for the winter. This information should come from what you think you can sell over your winter marketing period and how much of those crops you will harvest on a weekly basis. Because the crops take so long to regenerate over a winter, how many beds of each crop you have will determine how long you can be in production or how much you can sell each week.

For example, one winter season, I attended one Saturday farmers market all winter and sold to two restaurants. This marketing period started November 1 and ran to mid-March, when production for my main season started. The products I offered were pea shoots, sunflower shoots, radish microgreens, kale, spinach, lettuce mix and carrots.

Much of the product I grew was indoor microgreens, as I can easily grow those year round without hoping for a warm winter, but four field crops can be kept in the soil all winter and will survive even down to 5°F temperatures.

Winter farming preparations starts in midsummer, and the first winter crop I start to prepare is kale. For details on planting winter kale, see Part 10.

The next winter crop I plant are carrots. For details on planting winter carrots, see Part 10.

Because they are quick growing, spinach and lettuce are the last crops to be planted for the winter. With these two, it may be necessary to plant two successions. One group the third week of September, and the other the following week. As the days get shorter going into the fall, the time between crop maturity gets longer. What actually happens is that two plantings one week apart in mid-September may actually end up maturing many weeks or even a month or more apart. This is because the days get exponentially shorter after or near the fall equinox, and the evenings get a lot cooler. The further north you go, the more you notice how each day is visibly shorter than the last. Lettuce can be precarious as a winter crop if it gets too cold. I find that it has a tendency to fold over on itself as the temperature gets well below freezing, and then as it warms up during the day or when it's sunny, the lettuce will rot on itself. Spinach doesn't seem to have this problem because the leaves are a lot thicker and more robust.

### Overwintering Crops for an Early Spring Harvest

*Overwintering* is a process in which a cold-tolerant crop will be planted late in the season, with the intention of keeping it alive over the winter. The main crops I overwinter are spinach, kale, lettuce and carrots. There are two basic approaches with overwintering:

1. Plant a crop so that it will be mature by the time the cold arrives, and it will sit in a form of stasis for the winter.
2. Plant late in the season, just so the crop grows a very small amount before the cold; it will be ready to grow more vigorously in early spring the following year. This works for spinach and lettuce. This technique will not work in extremely cold climates like the Canadian prairies or northern regions. It will work in most of the USA. If temperatures are lower than 5°F for prolonged periods, than it may not work as well.

When planting for harvest the following spring, you start in the fall. For spinach and lettuce, see Part 10 for crop specs for each.

### Storage Crops

Storage crops are most often root crops that have the capacity to be stored over the winter in a controlled environment like a root cellar or walk-in cooler. This is nothing new: people have been doing this for thousands of years. Potatoes, carrots, beets, celeriac, onions and parsnips are common crops that farmers will store and market through the winter. On an urban farm, cold storage can be challenging, as you do need to have an expanded cooler area in order to be able to store a significant amount of crop to market for months at a time. With the specialized crop production that my farm focuses on, the only crops that I have the capacity to store are beets and carrots. In the past, I have planted many beds of beets and/or carrots during late summer in order to have a large harvest in early fall. When the carrots are harvested, they are washed and allowed to dry (so there is no standing water on them), and then they are stored in the cooler. Carrots become a little dry in the cooler, so it's good to wash them early; otherwise they are difficult to clean later. Beets can be harvested and stored dirty (they store better when they are dirty), and beets are easier to clean. If you don't have a root cellar, which is complicated for most urban farmers, you can put a small space heater in your walk-in cooler to keep its temperature between 35–39°F during the cold months.

For the size of my land base, I have found that it's more profitable for me to focus on quick-growing crops, with some overwintered crops for winter marketing, accompanied by solid microgreens production. On my tiny land base, I prefer to grow beets for storage and keep my carrots in the ground over winter. It's all about a diversified approach when it comes to the best season extension plan for your farm.

PART 9

# BASIC CROP PLANNING

I plan my farm very differently from most other market gardeners or small farmers. There are a few main reasons for this:

1. I'm on a very small land base, and I can't afford to be stuck to a rigid farm plan. If a certain crop was planned to occupy an area but that crop isn't selling, then I can't afford to leave that land occupied if it isn't making a profit.
2. I don't run a traditional CSA. Planning a farm around a CSA is simpler: you plan what you want to put in each box for each week of the season, then you work backwards to see how that relates to having crops ready for those box dates.
3. My market demand changes all the time. When you're dealing with restaurants, you need to be able to shift to changing demand. What was trendy one year can be old news the next. Because restaurants will often cater their menus to what their customers demand (as any good business person would do), I am at the whim of my restaurant customers. There are a lot of certainties—things that I know I can grow each year—but the overall crop lists do change from year to year, even during the middle of the season.

# Determine Your Outcome

In any farm plan, the first thing is to determine how much money you'd like to make. We know that areas in HR can conservatively generate $800 per 30-inch by 25-foot bed, and BR $400 per bed. So determine how much money you can make by quantifying the number of beds using these HR and/or BR benchmark figures. Let's look at some examples.

This chart provides six examples of farm sizes ranging from ¼ acre to ½ acre, all with a 30-week season. The column on the right represents the average weekly income to target. If your season is less than 30 weeks, simply multiply weeks in your season by the average weekly income to give a benchmark figure for total season outcome. For example, a 20-week season with example

**FIGURE 51.** Six Urban Farm Operations.

| Ex. # | Land in production | Total area in Square Feet | HR beds $800 (4 crops at $200 each) 30" × 25' (1' paths) 87.5 SF | BR beds $400 (2 crops at 200 each) 30" × 25' (1' paths) 87.5 SF | Gross profit per season | Average income per week |
|---|---|---|---|---|---|---|
| 1 | ¼ acre (All Quick crops) | 10,890 | 100 | | $80,000 | $2,667 |
| **2** | **¼ acre (Balanced crops)** | **10,890** | **60** | **40** | **$64,000** | **$2,133** |
| 3 | ⅓ acre (All Quick crops) | 14,375 | 130 | | $104,000 | $3,467 |
| 4 | ⅓ acre (Balanced crops) | 14,375 | 70 | 60 | $80,000 | $2,667 |
| 5 | ½ acre (All Quick crops) | 21,780 | 210 | | $168,000 | $5,600 |
| 6 | ½ acre (Balanced crops) | 21,780 | 110 | 100 | $128,000 | $4,267 |

#1 would be 20 multiplied by $2,667 for $53,340. This figure is what your target income for a 20-week season.

Keep in mind that these figures are goals to aim for, and there is no guarantee where you'll arrive. There are many variables in farming: weather, market demand and work ethic. All of these things combined can really determine your outcome after the fact. Use these numbers as targets to strive for.

In my first season, I did about 33% of these numbers, then in my second 65% and in my third 100%. Looking at example #1, if you are totally new to farming, a realistic target income might be 33% of $80,000 ($26,400)—a conservative number to achieve in your first year—then $52,000 in your next and $80,000 by your third.

**FIGURE 52.** Seasons Crops Become Available for Harvest

| Crop | Available for Harvest |
|---|---|
| Arugula | early spring |
| Basil | summer |
| Beet greens | summer |
| Beets | spring |
| Bok choy | spring |
| Carrots | summer |
| Cilantro | spring |
| Dill | spring |
| Salad turnips | spring |
| Kale | early spring |
| Lettuce | spring |
| Mustard | spring |
| Parsley | summer |
| Radishes | spring |
| Red Russian kale | early spring |
| Scallions | spring |
| Spinach | early spring |
| Summer squash | summer |
| Swiss chard | spring |
| Tatsoi | early spring |
| Tomatoes | summer |

In the rest of this part of *The Urban Farmer* I will use example #2 from this chart as our farm model.

### Crop Season Availability

Before we can decide what we're going to sell during the season, we first need to determine what will be available during each part of the season. For crop availability, I divide my season up into three harvest start periods: early spring, spring and summer. These are the main times when certain crops will become available, based on what I know can grow in my area. Mid-March through mid-April is early spring in my area. During this period, I am mostly harvesting indoor microgreens and overwintered crops like spinach, kale and sometimes lettuce and carrots that were planted in the fall. Spring season on my farm is from mid-April to mid-June. This is the period where most of my crop selection comes online: radishes, turnips, beets, early carrots, some herbs, new kale and many types of greens. The summer is the time when tomatoes, summer squash and lots of carrots start to produce high volumes.

Plan the basis for a season from a chart like this; from here you need to determine how many marketing weeks you're going to have. Keep in mind that your crop season availability will be different for your own bioregion. I would suggest finding similar kinds of information as locally as you can get.

Once you know which crops can be available at each point in the season in your own area, you can further break down that information into finer detail: to exactly when you want to have those crops and how much you want to sell each week. This is much like planning a CSA. To determine what to plant, count backwards from harvest date to planting date.

For example, if I want to have radish available for the 1st of July, I can simply take the average date to maturity (DTM) of that crop (28 days), and, counting 28 days backwards from July 1, I arrive at June 3. So my planting date for a July 1 harvest of radishes is June 3.

During the early spring however, the average DTM is not necessarily relevant because temperatures are much cooler and days are shorter. Using season extension techniques with high or low tunnels can help get the DTMs of your crops closer to the average. Otherwise, add a certain number of days to your expected DTM. This information is unique for every bioregion. Get to know farmers in your area who can give you a better idea of how day and night temperature changes effect crop cycles where you live. This concept is covered in greater detail in Chapter 40.

Once you understand this principle and its variables, basic crop planning is just planning what products you want to sell and how much you want to have available each week—then counting back from those dates to determine your planting dates. Having a clear understanding of your expected yields is also critical. That way you'll know how much to plant based on what you'd like to sell.

# The Base Plan

Refer to example #2 on the farm model chart (Figure 51). This farm is ¼ acre, targeting $64,000 income. This is broken down into 60 beds in HR and 40 in BR. The HR area will be dynamic, meaning the plan will consist of groups of crops we want to plant and sell on a weekly basis (such as radishes, turnips and various types of cut salad greens). I know there are certain things that I want to have available based on previous sales years. For example I know that I can sell up to 400 bunches of radish each week from April 1 to June 1. So, that means I need to have four to six beds of radishes to harvest each week for that period. I know I can sell 40 pounds of arugula each week, and that means I need to have around four beds to harvest each week. But I will not necessarily plant four beds a week because greens can be cut multiple times.

I also don't plan my Hi-Rotation areas because I can't guarantee how many cuts I will get from a bed of greens. Crop yield can change with the weather. Constantly monitoring crop development in all my HR areas, and making changes as I go, is the best way I've found to manage crop yield in HR areas. For example, I might have been planting four beds of arugula for a few weeks, and then I notice that I am getting more repeated cuts on those beds than I thought I would. So I might skip a week of planting. In most cases, this is how it goes.

During summer, however, I don't get as many repeated cuts because many of the greens I grow are spring crops that have a tendency to go to seed in hot weather. But again, the weather is always a little different each year; it's like the saying, "The only constant is change." I have accepted this, and it's partly why I don't plan my farm the way most farmers do. This farm planning system is not totally integral to success with an urban farm, but I believe it's been part of my success, as it has allowed me to adapt and change production on the fly.

The only aspects of my farm that I do plan in detail are the Primary Crops in my Bi-Rotation areas—mainly because they take so long to mature and need to be started in the nursery. In BR areas, the Steady Crop is the Primary Crop, and Secondary Crops are not planned. Why? Because demand for a certain crop may increase during summer, and I might decide to plant some Quick Crops in a BR area after the Primary Crop has been harvested. For example, by late summer when a Primary Steady Crop like kale gets pulled out of the ground as aphid pressure becomes too much to manage, I will sometimes put quick-growing crops in its place to get me through to the fall. That could be something like lettuce, fall beets and carrots. But it's not planned but approached in the same way I manage HR areas.

One crop that I plan in detail is carrots. I don't always know where they are going to go because I plant carrots in HR areas most of the time. But I do know how many beds of carrots I want to plant and how often. Since carrots are a longer-season crop, it's difficult to scale production up quickly based on changing market demand. I approach planting carrots with a determined outcome. I decide how much I want to have available each week for the season and plant from there. For example, I plan to have one bed of carrots available for harvest each week from the time they are in season. Usually, that is early June or late May.

### Succession Planting

Succession planting can be one of the most difficult aspects of farming because it involves understanding your market demand, changes in day length and temperature from season to season, what you can expect your crops to yield and how long you expect to harvest them. A farmer must make succession plantings in order to achieve a consistent harvest from his/her farm. In order to achieve a steady income from your farm, you must consistently plant and harvest throughout the season.

Before you can start to succession plant, you need to have an idea of what demand there is for a particular crop (how much you think you can sell on a weekly basis). Once you have an understanding of how much you can sell, then you need to look at how much yield you need—and then, you will know how much you can plant. The part that makes it complicated is when do you plant, and how do the changing day lengths and temperature variations during the year affect when those crops will be ready?

Let's say that, between a weekly farmers market and a few restaurants, I think that I can sell around 150 bunches of radishes per week. Based on my crop profile information, I know that I would need two 25-foot beds each week to harvest this much successfully. If I need to have those radishes ready by May 1, then I can look in the seed catalogue and see that radishes have a date to maturity (DTM) of around 30 days. If this were actually true, then I could plant radishes 30 days before May 1, and then they would be ready. It would also mean that I could plant every week for a continual harvest thereafter. This is also not true.

The further north or south you go from the equator, the more extremely day length changes from season to season. If you keep track of your own crop information by recording planting and harvesting dates, after one season of farming you will have a much clearer idea of how varied day lengths in your area affect succession planting.

In downtown Kelowna, BC, (zone 6B) if I were to direct seed radish in the field on April 1, my expected date to maturity is 41 days. If I were to plant again two weeks after, on April 14, my DTM would now be 32 days. Day length gets exponentially longer during this phase of the year, and the temperatures are also rising, thus shortening the date to maturity for crops. In my first year farming, I made the mistake of planting every week during this early spring period. Crops that I planted April 1 and then again on April 8, ended up both being mature at the same time.

What I have learned is to plant by *crop development*, not by any particular schedule. Instead of deciding to plant a crop like radish at fixed days, I plant the next succession when the first one has reached a certain stage of growth. For example, when I plant radishes on April 1, I won't plant again until the first planting has emerged from the soil and sprouted two true leaves. Then, I know that there is enough lead time to start the next crop so that they won't mature at the same time. I will also plant two weeks' worth of harvest at once instead of planting one week at a time. When the weather is cooler, radishes can stay in the ground longer without maturing too quickly. During spring, I plant more beds at once, with wider intervals of time between plantings.

Once warm weather begins and the day and nighttime temperatures stabilize, I then move into a consistent weekly schedule of planting. At this point, I no longer plant by crop development. In the case of radishes, once we're past the second week of May, their DTM becomes more consistent. At this point, I will plant only one week's worth of harvest at a time because I will need to get them out of the ground within a week. Radishes can grow so fast in the summer that they can turn woody in a matter of days. Because of this, they must be harvested fast. During summer, I will plant small amounts at tighter intervals, where in the early spring it is the opposite (planting large amounts at wider intervals).

### Cut and Come Again Greens

I refer to greens like arugula, mustard, lettuce and spinach as *Cut and Come Again Crops*. They can be cut multiple times as the same plant regenerates repeatedly two weeks or less after the first harvest of leaves. This regeneration can greatly boost your yields, because if you plant every week consistently you will have new greens becoming available at the same time as second and third cuts from previous plantings. If you understand this process, you can use it to your advantage. It took me a few years to figure this out, as there were many times throughout my first seasons when I was stuck with too many greens to sell or I was turning in beds because I knew I couldn't sell the product. You want to avoid that altogether.

There are several succession planting strategies for Cut and Come Again Crops that I use. Depending on how many greens I think I can sell, each will offer a slightly different yield and planting schedule. Figure 53 sets out my strategy for planting arugula and spinach. Figure 54 shows how I plant loose lettuce. Figure 55 describes plantings for mustard and Red Russian kale.

Each crop can be cut at least three times. Red Russian kale and lettuce can be cut four times or more. Greens don't always regenerate at the same two-week intervals, and there will be variability for your own climate and time of the season, but these tables illustrate the cycles of the regeneration process and how it pertains to crop planning.

In Figure 53 our planting dates for arugula and/or spinach are on the top row: May 4, May 11, May 18 and so on. Our harvest dates for each of those plantings are in the columns below, and the number in each box before the date represents the sequence of harvest. Starting from the beginning, let's say we plant one bed of arugula per week, and it will yield ten pounds. If we plant May 4, and each week after, our first harvest will be May 28 for ten pounds, and our next week's harvest will be June 4 from a May 11 planting. Our third harvest in sequence is June 11. Notice how it is the first cut from the May 18 planting as well as the second cut from the May 4 planting. This week we are harvesting 20 pounds because we're cutting from two plantings. This is how succession planting will start to overlap with previous plantings. The first succession occurs on June 11, where the first cut from May

**FIGURE 53.** Arugula and Spinach Succession Plan.

| Planting Dates | May 4 | May 11 | May 18 | May 25 | Jun 1 | Jun 8 | Jun 15 | Jun 22 | Continues... |
|---|---|---|---|---|---|---|---|---|---|
| Harvest 1st cut | #1<br>May 28 | #2<br>June 4 | #3<br>June 11 | #4<br>June 18 | #5<br>June 25 | #6<br>July 2 | #7<br>July 9 | #8<br>July 16 | ... |
| Harvest 2nd cut | #3<br>June 11 | #4<br>June 18 | #5<br>June 25 | #6<br>July 2 | #7<br>July 9 | #8<br>July 16 | #9<br>July 23 | #10<br>July 30 | ... |
| Harvest 3rd cut | #5<br>June 25 | #6<br>July 2 | #7<br>July 9 | #8<br>July 16 | #9<br>July 23 | #10<br>July 30 | #11<br>Aug 6 | #12<br>Aug 13 | ... |

18 and second cut from May 4 overlap. The second succession is on June 18. Notice how on June 25 we reach a point where there are three crop successions being harvested at once, which would be 30 pounds of arugula or spinach. This is the critical point where your production from these crops will be at its peak. From this point on you will be cropping from three plantings at once.

My second crop succession technique is planting two weeks on and one week off. This has been my primary schedule for loose-leaf lettuce, and I will always get at least three cuts and sometimes four. During the summer months, I will most likely just get three because heat will stress the lettuce and cause it to bolt. I plant May 4 and May 11, and then skip the next week. I resume May 25 and June 1, then skip. Peak production on this schedule occurs by June 18, where I am harvesting the first cut of my May 25 planting and the second cut of the May 11 planting. The purpose of this schedule is to harvest only two plantings at

**FIGURE 54.** Loose-Leaf Lettuce Succession Plan.

| Planting Dates | May 4 | May 11 | Skip week | May 25 | Jun 1 | Skip week | Jun 15 | Jun 22 | Continues... |
|---|---|---|---|---|---|---|---|---|---|
| Harvest 1st cut | #1 May 28 | #2 June 4 | | #4 June 18 | #5 June 25 | | #7 July 9 | #8 July 16 | ... |
| Harvest 2nd cut | #3 June 11 | #4 June 18 | | #6 July 2 | #7 July 9 | | #9 July 23 | #10 July 30 | ... |
| Harvest 3rd cut | #5 June 25 | #6 July 2 | | #8 July 16 | #9 July 23 | | #11 Aug 6 | #12 Aug 13 | ... |
| Harvest 4th cut | #7 July 9 | #8 July 16 | | #10 July 30 | #11 Aug 6 | | #13 Aug 20 | #14 Aug 27 | ... |

**FIGURE 55.** Mustard/Red Russian Succession Plan.

| Planting Dates | May 4 | Skip week | Skip week | May 25 | Skip week | Skip week | Jun 15 |
|---|---|---|---|---|---|---|---|
| Harvest 1st cut | #1 May 28 | | | #4 June 18 | | | #7 July 9 |
| Harvest 2nd cut | #2 June 4 | | | #5 June 25 | | | #8 July 16 |
| Harvest 3rd cut | #3 June 11 | | | #6 July 2 | | | #9 July 23 |
| Harvest 4th cut | #4 June 18 | | | #7 July 9 | | | #10 July 30 |

a time, unlike technique #1 in which I harvest three.

Figure 55 is a schedule I use to plant Red Russian kale and mustard greens. During my main season, these greens regenerate in one week and will deliver at least three cuts.

This schedule offers consistent production with little or no overlap; if there is a fourth cut, the fourth cut of the first planting and the first cut of the third planting will overlap. If you experience a very hot summer, you may need to slightly modify this plan. During July and August, I would suggest tightening your planting intervals to accommodate fewer repeated cuts. You might have to move to weekly or two-week intervals.

PART 10

# CROPS FOR THE URBAN FARMER

This part contains a comprehensive list of the main crops I grow on my farm. Since 2010, I have grown around 80 varieties of annual vegetables, but I have found these to be the most lucrative and realistic for an urban farmer on a small land base. In Chapter 5 I outlined the five main points that make a crop suitable for growing in the city; I call this *crop value rating* (*CVR*). Keep in mind that if your goal is to generate as much income as possible on a small land base, you must focus your crop portfolio on higher-value and quick-growing crops.

**Crop Type:** Quick or Steady
**Seasons in Production:** Months you can harvest the crop
**Crop Value Rating (CVR):** How many CVR characteristics this crop possesses
**Planting Specs:** Details of the density and seed used in a bed
**Varieties:** Preferred varieties to grow
**DTM:** Average date to maturity from direct seeding and transplanting
**Average Yield Per Bed:** The total yield that you can expect to harvest, including multiple pickings or cuts
**Average Gross Profit Per Bed:** How much you can expect to make per bed

## Arugula

Arugula is one of the most popular crops I grow for chefs and the farmers market. It is very trendy in the culinary industry. I sell it on its own and put it into my salad mixes. My favorite type of arugula to grow is a serrated leaf, commonly referred to as Wild Arugula.

**Production:** Arugula grows strongly and yields high volume in the spring and fall, but will decrease during the hot months of summer. I will start spring planting as early as the first week of March for greenhouse production and the third week of March for field production under poly low tunnels. After this point, I'm succession planting based on crop development, which is usually two plantings about two weeks apart. From May until September, I'm planting every single week. As the days get longer I'll increase the frequency of plantings but decrease the number of beds that I plant. In the springtime with arugula, I will sometimes get over three cuts per bed. But in the summer, I only get two cuts per bed, in some cases just one. Arugula is a prime crop for urban production, as it scores 5/5 on the CVR. Arugula is a must for urban farmers.

**Harvesting:** I try to grow a baby crop so I can cut it short and get as much of the shape of the leaf as possible without too much stem. During the summer months, this means cutting it really young and getting a smaller yield. In hot weather, arugula needs to be cut early because it will bolt very quickly. This sometimes means harvesting it before your main harvesting day. By mid-May, a freshly cut bed of arugula will regenerate in one week. If I cut it on a Thursday, it should be ready to cut again the next Thursday. During the cooler months it regenerates

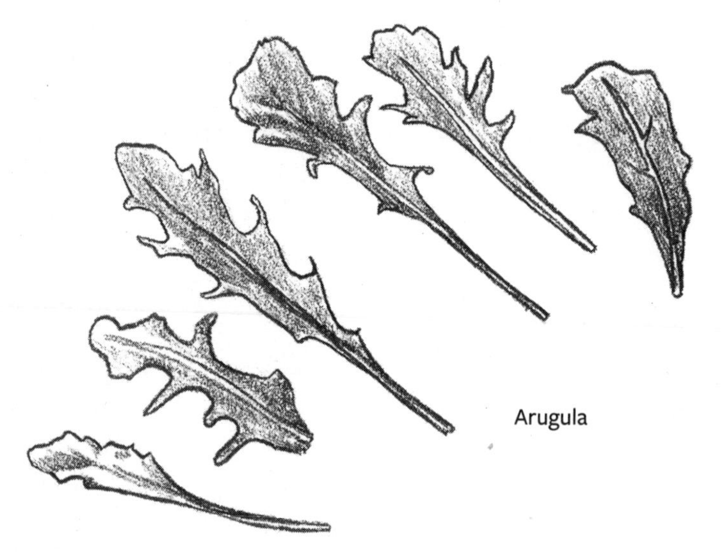

Arugula

more slowly, in around ten days to two weeks. To maintain steady weekly production I harvest on a Monday one week, then on Thursday or Friday the next week. This way you give the plants more days to regrow. Once you have multiple beds going, you have fewer gaps in production from different succession plantings. For cutting, we use the Quick Cut Greens Harvester or very sharp serrated steak knives for hand harvesting.

**Crop Type:** Quick
**Seasons in production:** April through October
**CVR:** 5/5 (short DTM, high yield, high price, long season, popular)
**Planting Specs:** Direct seed at 9 rows; Seed use: .41 oz; Seeder: Jang YYJ24 roller
**Varieties:** Rocket, Voyager, Sylvietta
**DTM:** In spring 35 days; in summer 21 days
**Average yield per 25-foot bed:** 30 pounds from up to three cuts
**Average gross profit per bed:** $300 at $10 per pound

### Auxiliary Greens

Auxiliary greens are grown exclusively for the various salad mixes I offer and nothing else. Often they are grown in short beds, at inconsistent frequencies. They are not a priority crop, so I plant them in places where I have extra space. The main greens I grow are mustard, tatsoi and beet greens. These greens add a bit more color and texture to salad mix.

**Production:** Tatsoi is a high-yielding early green, and it can deliver many cuts week after week. It's one of the first crops I sow in the greenhouse, and it will regenerate every week from the first cut in early April for four to six weeks. As soon as the weather gets warm, it will bolt very quickly. I usually just do one sowing of tatsoi in the greenhouse in mid-March, then one more sowing in the field in early April. Mustard has a little bit more versatility, as I can keep planting it throughout the summer. The first greenhouse and field sowings will deliver many cuts but into summer, mustard is good for only a maximum of two cuts. Beet greens are my main summer auxiliary green. Beets like warm temperatures and will deliver many cuts week after week. The first sowing is in June, as I'm looking for optimal temperatures. Beets take too long to germinate if the weather is cool, so I wait until it's warm.

**Harvesting:** For harvesting all these greens, I prefer the Quick Cut Greens Harvester but can cut by hand as well. Mustard will get taller after each cut, so you have to cut it higher to get less stem and more leaf. Tatsoi and beet greens stay pretty low to the ground, so harvesting them over time is easy. Make sure to harvest beet greens when they are three inches tall. They can get too big fast, especially when it's hot.

**Crop Type:** Quick
**Seasons in production:** April through October
**CVR:** 5/5 (short DTM, high yield, high price, long season, popular)

**Planting Specs:**

**Mustard, Mizuna, Tatsoi:** Direct seed at 9 rows. Seed use: .78 oz. Seeder: Jang X24 roller

**Beet Greens:** Direct seed at 7 rows. Seed use: 2.9 oz. Earthway seeder, Chard Plate

**Varieties:**

**Mustard:** Scarlet frills, Mizuna

**Beet Greens:** Bull's Blood, Early Wonder

**DTM:** In spring 35 days, in summer 21 days

**Average yield per 25-foot bed:** 20 pounds from up to three cuts; average yield is six pounds per cut

**Average gross profit per bed:** $200 at $10 per pound

### Basil

You can command a high price for basil if you can have it ready by mid-spring, but when everyone else has it in the summer the price goes to less than half of what you get for early sales. In my experience, basil is a popular crop for my farmers markets, but less popular with chefs. Once or twice a season, chefs will buy large amounts for making pesto. And, during the later months of summer, farmers market customers will buy large amounts in bulk to make pesto for themselves during the canning season. Most of the time, I do only two plantings of basil. The early crop, started in the nursery early March, is transplanted into the greenhouse by early April. I will get small

Pinch the center stem after the first two nodes have formed, then the center of each branch after the first picking. Pinching back causes the plant to keep growing from its base; it forms a little bush.

pickings in late May, and at this point I sell basil in small two-ounce bags for $3. That's $24 per pound. The second planting will be started in June for a summer to fall harvest. At this point I can only get $8 per pound. So, I prioritize the early planting and try to have a lot available for times of scarcity.

**Production:** I plant at six-inch centers and am looking for many pickings without a lot of stem in high frequency. I will often interplant basil amongst my tomatoes: one or two rows at six-inch spacing, about eight inches from the tomato row. The tomatoes won't shade out the basil until later in the summer; at that point, you could have another crop going somewhere else.

**Harvesting:** I find the first harvest will be very light, but will double after each harvest until a point in the summer when you're picking 20 pounds per bed every week. At the point where I'm picking this much, and the price has gone down so low, I will sometimes abandon it for the rest of summer, to get something else in of higher value. My approach has always been to never compete for low prices.

**Crop Type:** Steady

**Seasons in production:** June through September

**CVR:** 3/5 (high yield, high price, popular)

**Planting Specs:** Transplanted at six-inch centers in the bed; 2–3 seeds per cell; four rows per bed

**Varieties:** Sweet, Italian

**DTM:** 76 days from seed; 24 day from transplanting

**Average yield per 25-foot bed:** 25 pounds; can be picked multiple times

**Average gross profit per bed:** $250 at $10 per pound

### Beets

Beets are a staple crop on my farm. I primarily sell baby beets to chefs and will also bring them to market until June. By this time, every other farmer has large, full-size beets, and I'd rather not compete in that saturated market. From my end, growing full-size beets means I'm not getting the best price because they need to stay in the ground too long; on an urban farm, I'm better off turning that ground over to get more crops in than letting slow-growing crops sit to full maturity. Most chefs prefer a golf-ball-size beet with one inch of green left on. Beets are often displayed on a plate lightly peeled and split in half with a little bit of green left on. Golden beets are a favorite among chefs I know because they don't bleed their color like red beets do. Chioggia beets are similar, but I find I have less demand for them so I don't grow them that often.

**Production:** Since I'm targeting a smaller beet, I plant them at high density. I plant four rows in a 30-inch bed; I will transplant my first crop and direct seed the rest. I do around four plantings of beets per year. The first is started in my nursery around early March, and the last planting is direct seeded on the first week of August: that will be a fall/winter crop. I plant enough to last until late November. Beets don't overwinter that well, especially when we get cold winters (below 5°F). But there have been times where I've kept beets going all winter long under poly low tunnels. If your winters are consistently above freezing, it is possible to keep them alive in the ground all winter.

**Harvesting:** When harvesting beets, I am thinning them to pull the largest ones first, which frees up space for the smaller ones to grow bigger by the next week. I may harvest one bed for three weeks at a time. I pack harvested beets into perforated totes loosely, then bunch them for market as they are washed. Unless otherwise requested, we sell beets to chefs with all but two inches of the tops removed. We keep the tops on for a nice display when sold at market.

**Crop Type:** Steady

**Seasons in production:** June through November

**CVR:** 4/5 (high yield, high price, long harvest, popular)

**Planting Specs:** Four rows per bed; Nursery planting 1–3 seeds per cell, transplanted three inches apart in bed

**Varieties:** Red ace, Touchstone Golden, Chioggia

**DTM:** Spring, 80 days from nursery seeding; Summer, 50 days from direct seeding

**Average yield per 25-foot bed:** 100 bunches, multiple pickings

**Average gross profit per bed:** $300 at $3 per bunch

Beets

### Bok Choy

Bok Choy is a great early season crop and can be good for fall production as well. I sell small bunches of choy at the market during the early spring, but find customers lose interest in it by late spring. At this point, I keep growing it for chefs. Fine restaurants are usually looking for the smallest type you can grow. Chefs like to break off the whole leaf and stem from the base of the plant, sauté or steam that so that it looks whole. My favorite variety is called Shiro, a small and compact variety with a very short DTM. The biggest challenge with bok choy is to keep it from bolting, so I don't really grow it past the end of May. It grows well in cool soil and tastes better before warmer weather stresses the plants.

**Production:** I always start bok choy as transplants, even for a fall crop. With the hot and dry summers in my area, I can't get bok choy to germinate even in early September, so a fall crop is better started in plugs under shade and transplanted out in September. The months I market choy are April and May, from greenhouses or poly low tunnels. I set October-November as transplants into the field, then cover with poly low tunnels by early October, at which point bok choy will hold its size as the weather gets cooler and the days shorter.

**Harvesting:** I use a simple small knife to cut the root just under the plant and carefully pack the plants upright. Bok choy is fast to harvest, but you must be careful to wash it. You need to handle it gently, especially the small varieties, as the leaves have a tendency to break easily.

Bok Choy

**Crop Type:** Quick
**Seasons in production:** April through May, October through November.
**CVR:** 3/5 (short DTM, high yield, high price)
**Planting Specs:** Four rows per bed; Nursery planting 1–3 seeds per cell, transplanted three inches apart in bed
**Varieties:** Shiro for small heads, Joi Choi for medium heads
**DTM:** Spring time, 80 days from nursery seeding; summer, 50 days from direct seeding
**Average yield per 25-foot bed:** 50 pounds
**Average gross profit per bed:** $250 at $5 per pound

## Bunching Herbs

Bunching herbs can be a highly lucrative crop, almost comparable to microgreens on a square foot basis. The challenge with them, is that nobody ever needs a lot at once. There are always exceptions, but I'm never selling hundreds of bunches of herbs at a time. Because I've never been able to find a huge market for them, I grow then in half-sized beds and produce 40 or so bunches each per week, to offer variety to my chefs and farmers market customers. The three most popular bunched herbs are cilantro, parsley and baby dill. Cilantro scores a 5/5 on the CVR, where the other two still are 4/5, mainly because they have a slightly shorter season.

**Production:** For all three of these herbs, I plant in 6- to 12-foot beds at a time.

**Harvesting:** I bunch all herbs in the field and eyeball the weight. They mostly end up weighing about two ounces per bunch. I use a very sharp knife to harvest them, and keep a small bin of elastic bands with me as I go.

### *Cilantro*

This is a popular herb for chefs during most of the season, and you want to make sure you have a lot at market for late summer, when everyone is canning salsa. Cilantro can be grown most of the season, but will slow down a little bit during the really hot months; at this point I can still plant it, but I will get fewer cuts. So, just as I do with arugula, I'll plant smaller batches more frequently.

Cilantro

### *Parsley*

This is mostly a summer to fall herb. It doesn't do to well in the early spring, but if you start it as transplants, you can get an earlier crop. I wait to direct seed it most of the time, around mid-May, and will do a few more plantings until late summer. Parsley can winter over if it's covered.

### *Baby Dill*

I grow a very small amount of dill, and never let it get big. I'm targeting a small leafy bunch that is no more than six inches from base to top. The seed is expensive, but chefs love it. Dill can be challenging with aphids as you get into summer, so for this reason I grow it only in the spring.

**Crop Type:** Quick

**Seasons in production:**

- **Cilantro:** mid-May through October
- **Parsley:** mid-June through October
- **Baby Dill:** mid-April through May

**CVR:** Varied

**Planting Specs:** Varied

**Varieties:** Cilantro: Calypso; Parsley: Italian; Dill: Fern Leaf

**DTM:** Cilantro: 30 days; Parsley: 70 days; Baby Dill: 55 days

**Average yield per bed:**

- **Cilantro:** 250 bunches per 12-foot bed
- **Parsley:** 235 bunches per 12-foot bed
- **Baby Dill:** 200 bunches per 12-foot bed

**Average gross profit per bed:**

- **Cilantro:** $500
- **Parsley:** $470
- **Baby Dill:** $400

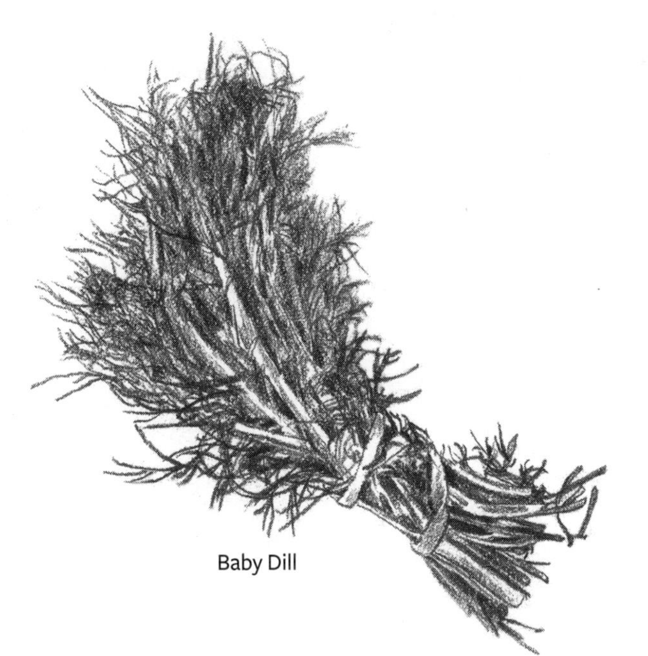

Baby Dill

### Carrots

Carrots are a classic crop. The best thing about them is that everyone loves them. We grow a lot of carrots for chefs and restaurants, but each point of sale is looking for different characteristics. Chefs are looking for almost a baby carrot that is just big enough to not have a core formed in the center. These carrots are about the width of a thumb, and have the most amount of flavor. Rainbow varieties are popular, but the standard orange carrot is still the most popular.

Most high-end restaurants are looking for a carrot with just a ½ inch of green on the top, for the way they are displayed on the plate. Chefs will also take super small carrots (that are basically thinnings) for special events. In this case, I'll plant nine rows in a 30-inch bed, and double seed each row. For markets, I'm targeting a slightly larger carrot—not a monster for juicing, but one big enough to form a little bit of a core down the middle.

**Production:** I plant carrots five times per season. The first is under poly low tunnels or greenhouses the first week of April, then every month after that, with the last being an overwintered crop, seeded August 1. In my climate, I can overwinter carrots and pull them from the ground all winter. The key is to keep the soil dry during the fall. Winter carrots are fully mature before the first week of October, then they get covered with poly low tunnels and no longer get watered. If you can keep off the winter precipitation, then when the weather drops to freezing, your soil will not freeze. This is critical to storing carrots in the ground. For this to work, you will also need to have soil with good drainage. If the carrot bed is in a low-lying area where water settles, this will

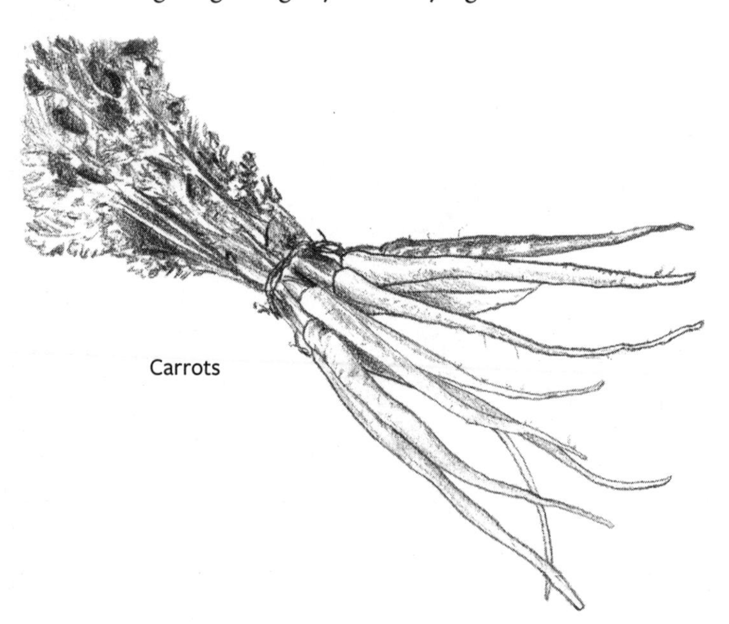

Carrots

not work, as the carrots will rot as soon as the soil thaws after a freeze.

**Harvesting:** I use a standard garden fork to loosen the soil, one row at a time. I start with the outside row of a bed first, loosen the soil about an inch or two from the top. If the carrots are going to market, I snap the tops off; if they're going to chefs, I will twist off the greens so that I leave a few inches on. Chefs often prefer to have one inch of green on the carrot. Sometime when the greens of the carrots get too long, they can crowd out your walkways, which can make it difficult to navigate the beds. To make it easier to walk in the beds you can cut the greens off before harvesting. Take a sharp knife and cut off the greens while the carrots are in the bed. Cut them about one to two inches off the ground, then proceed to fork them out.

**Crop Type:** Steady

**Seasons in production:** June through March

**CVR:** 4/5 (high yield, high price, long harvest, popular)

**Planting Specs:** 7 rows for baby carrots (¼ oz seed), 5 rows for full size and OW crop (.17 oz seed); Jang seeder XY24 roller

**Varieties:** Mokum, Rainbow, Purple Haze, Chantenay for micros

**DTM:** spring 68 days; summer 54 days

**Average yield per bed:** 60 pounds baby to medium sizes; 75 pounds full size

**Average gross profit per bed:** $240 baby to medium size at $40 per pound; $225 full size at $3 per pound

## Kale

Kale has become such a trendy crop in recent years that it's getting harder to find enough seed for large area plantings. I have seen some seed companies selling seed for $1,000 per pound. Known for its many health benefits, it's also a great urban crop, scoring a 4/5 on the CVR scale. I grow a lot of kale for market, distributors and chefs. I can grow kale almost year-round in my climate. Aphid pressure gets so high starting into July that I'd rather not deal with it during those hot six weeks. I start a second crop in plugs the second week of July, and set them out mid-August. I usually start to get some nice harvests by mid-September. This late planting of kale can be easily overwintered. If you plant enough to last you a winter, you want to have it fully mature and standing at least one foot off the ground before the first major frost. By picking the bottom leaves first, you train the plant upward, which will keep it high off the ground. In my climate, kale can survive six inches of snow on the ground and be ready to harvest again by mid-March. If I do a late summer planting in my greenhouse, I can harvest all winter long as well. Even though kale is not a super-high-value crop, because it can be harvested so many times and yields highly per square foot, it's a very worthwhile crop for an urban grower.

**Production:** I do only two plantings of kale per year, one in the early spring (where transplants are set out the first week of

April) and one in the fall (where transplants are set out the first week of August).
**Harvesting:** If the plant is fully mature and strong, you can harvest huge handfuls at once by pushing your hands downward with the leaves toward the ground. When the plants are young, you have to be more gentle with harvesting. The more you crop it, the stronger kale gets.
**Crop Type:** Steady
**Seasons in production:** All year except mid-August through mid-September
**CVR:** 4/5 (short DTM, high yield, long harvest, popular)
**Planting Specs:** Transplants at 10" centers, 3 rows per bed
**Varieties:** Winterbor, Darkibor, Redbor and Toscanna
**DTM:** Spring 74 days from seed, 35 day from transplanting
**Average yield per bed:** 115 pounds, multiple harvests
**Average gross profit per bed:** $575 at $5 per pound

### Lettuce

Lettuce for salad mixes is the most staple crop on my farm. I sell hundreds of pounds a week to chefs and will sell a hundred or more bags at the farmers market each week. When it comes to lettuce for restaurants, the chefs want consistency and a small-size leaf to use in mixed salad.
**Production:** Predominantly, I have direct seeded loose lettuce into the field, and it functions as any other Cut and Come

Kale is best harvested by hand, by snapping the leaves from the base of the plant downward.

Lettuce

Again Green like arugula or spinach. For direct seeding, I plant lettuce almost weekly during the main season. I typically grow between five and seven varieties of loose lettuce that do well in my climate, and all have the same date to maturity. This way, I can mix the seed together and plant them all in the same bed. The varieties have changed over the years, and they will change even through the season. Make sure you're growing varieties that all have DTMs that are within a few days of each other so the harvest is consistent.

In recent years, I have experimented with growing heads for leaf lettuce. For this approach, I grow a variety called Salanova®. We plant them as if they were grown for normal head lettuce, but we harvest them by hand and they grow back for many cuts. I have found the quality to be very consistent, and chefs love it. This type of lettuce grows differently than any other lettuce I've seen: the leaves don't get bigger as the head matures, it just grows more leaves. The great thing about this is that you don't have to worry about harvesting the beds on an exact schedule, which offers a lot of flexibility for times when you might producing too much. We use landscape fabric with holes burned in at six-inch centers, then transplant the lettuce in the holes.

**Harvesting:** To harvest lettuce by hand, we use a sharp serrated steak knife. Straddle the bed and move backwards as you harvest. The bin you harvest into will move on the bed where you have already cut. It doesn't damage the crop in any way. If you are right-handed, you hold the knife in your right hand and use your left to grab the greens. Harvest two feet of a row at a time, starting on your right side. Take your knife and cut one foot of greens in the row towards you, as your left hand catches them, and then move your hand to your harvest bin.

Once you get the hang of it, the most efficient way to harvest is to get as much crop into your hand as possible before you dump them in your bin. Starting with the row on your right, as you cut two feet towards you, you will cut the next rows to the left, until you have harvested the entire two-foot area in front of you. Now, move back two steps, and repeat the process. Once you get good at this, one bed can be harvested in 20 minutes or less. See photo #21 in the photo insert section.

**Crop Type:** Quick
**Seasons in production:** April through November
**CVR:** 5/5 (short DTM, high yield, high price, long harvest, popular)
**Planting Specs:**
    **Direct seeded:** Nine rows per bed. 1 oz of seed per bed. Jang roller F24
    **Mini-heads:** 4 rows, 6 inches apart, as transplants
**Varieties:** Salanova®, Red and green salad bowl, Oak leaf, Rouge D'hiver, Tango and Red sails
**DTM:** Spring 45 days for baby greens, 21 days in the summer
**Average yield per bed:** 40 pounds per bed from multiple cuts (up to four)
**Average gross profit per bed:** $400 per bed at $10 per pound

### Microgreens

The production potential with microgreens can be huge. With a small amount of space one can generate thousands of dollars of extra income a week with very little physical effort. The key is being able to sell them; otherwise they can be very costly. I sell a lot of micros to chefs, farmers markets and distributors, and they have added a lot of variety to my farm and allowed me to scale up very quickly. Growing indoors can be challenging because you must carefully

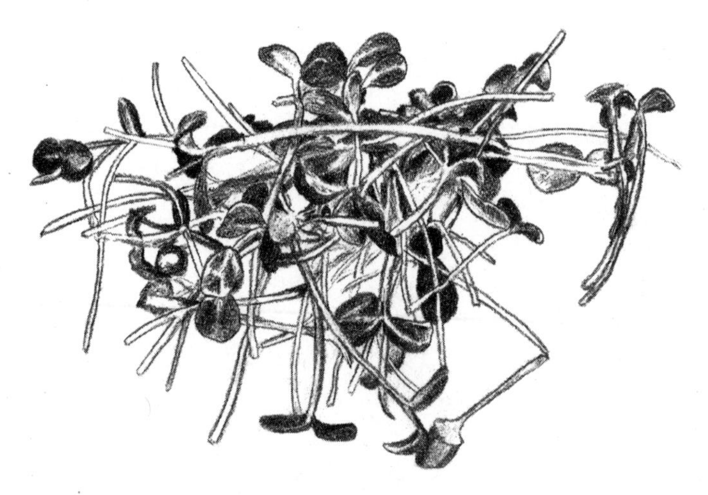

Sunflower Shoots

monitor humidity, airflow and temperature. Slight changes can cause fungus problems that can wipe our your harvest very quickly. Microgreens score a 4/5 on the CVR only because they are such a niche crop that it can be a challenge to market large volumes.

**Production:** The easiest way to get into micro-greens is to grow them during the main farming season in your greenhouse or outdoors. They are far less risky then. They can quickly provide great cash flow to your farm. See Chapter 37 for details on growing.

**Harvesting:** When harvesting micros in flats, I use a very sharp chef's knife. I carefully grab a handful of micros on the flat with my left hand, then cut with my right. In one cut, I'm cutting about ⅙ of the flat. In about six handfuls, I am cleanly harvesting the flat. I cut as low to the flat as I can, to get as much yield as possible. I primarily use one-inch-deep germination flats for this reason. This way, I'm cutting almost to the base of the soil.

When harvesting micros from the field, it's a similar technique, but I don't use the chef's knife. Instead I use a clean box cutter. Keep it washed and sterilized. I use a box cutter because I don't want to carry a sharp chef's knife around outside or tote it from plot to plot.

**Crop Type:** Quick

**Seasons in production:** All

**CVR:** 4/5 (short DTM, high yield, high price, long season)

**Planting Specs:**

- **Seed per flat:** Pea: 11.72 oz; Sun: 6 oz; Radish: 2 oz
- **Seed per 30-inch by 6-foot bed:** equivalent to 10 flats

**Varieties:** Black oil sunflower, Speckled Pea, China Rose Radish

**DTM:** 10–14 days

**Average yield per flat:** Sunflower: 1.5 lbs; Pea: 1.25 lbs; Radish: 1 lb

**Average gross profit per flat:** Sunflower: $22.50 at $15/lb; Pea: $18.75 at $15/lb; Radish: $20 at $20/lb

Pea Shoots

### Pattypan Squash and Zucchini

Summer squash is one of those crops that I grow largely for chefs. I will take them to market a handful of times during the season, but for the most part they are a staple amongst high-end restaurants. Much like turnips and beets, we're targeting another small vegetable. Pattypan optimum size for chefs is ping-pong-ball size with some stem. They want to be able to split them in half or serve them whole. With zucchini, I pick them at between three and four inches long. I always try my best to pick them small so I can get $4 per pound. But sometimes you'll miss some, in which case I charge $2.50 per pound for medium sizes and $1.50 per pound for large sizes. I take the medium and large ones to market, and the smalls are exclusively for chefs.

**Production:** Pattypans and zucchini are considered a Steady Crop and will be planted into BR areas. The plants will be in the ground from mid-May until mid-October (or the first frost). Summer squash is the Primary Crop if it's planted in BR areas, and in most North American climates there's an opportunity to plant a Secondary Crop before the squash plants are set into the ground from transplants. Crops like radishes, arugula and spinach make good Secondary Crops to plant before the Primary (in this case pattypans) go in the ground. Once transplanted, it's possible to interplant another bed of Quick Crops between the rows (review Part 8 on interplanting).

**Harvesting:** If pattypans or zucchini stay on the plant too long, they'll be too big and you won't get the best price for them. When the weather is hot, I like to pick them very early in the morning because I wear pants, long sleeves and gloves because the plants are a little thorny and can cause skin irritation. When picking pattypans, you need to move the plants around carefully to expose the fruits, which are often at the base of the plant and can be hidden. When you're picking, try to gather as many in your hands as you can before moving back to your bas-

Pattypans

ket. I snap the flowers off with my thumbs right after I pick them, and then keep a few fruits in the palm of my hands as my fingers continue to pick. Picking zucchini is similar except that it's good to have a small knife or pair of pruning scissors to cut the fruit from the plant, especially when harvesting baby fruits. If you don't cut at the stem, you can break part of the fruit if you try to twist it off.

**Crop Type:** Steady
**Seasons in production:** July through September
**CVR:** 3/5 (short DTM, high yield, popular)
**Planting Specs:** Single row, 18" apart.
**Varieties:** Patty pans: Sun burst; Zucchini: Raven, Golden Delight
**DTM:** 60 days from seed; 42 days from transplanting
**Average yield per bed:** 80 pounds per bed
**Average gross profit per bed:** $320 per bed at $4/pound

### Radishes

Believe it or not, radishes are one of my best crops. I barely sell any at the farmers market, but I sell hundreds of pounds per week to restaurants. For chefs they are a versatile vegetable: they can be diced or shredded; they are also commonly lightly braised, roasted and pickled. I deal with multiple chefs who will buy 25–50 pounds every single week. The best part is that they want them all season, and this is a crop that grows all season long. It does particularly well on the shoulder seasons. It's best to harvest them all when they're ready; the DTM on some varieties is as fast as 21 days, and if they stay in the ground too long during hot weather, they will get woody and hollow. Radishes can also be prone to some pest problems like cabbage root maggot. To combat this, I use an 80-gram insect netting to cover the crop. The netting lets light, water and air through, but it keeps the fly of the root maggot from landing on the foliage and establishing itself.

**Production:** In a 30-inch bed, I plant radishes at seven rows, but in the summer, I will sometimes plant one or two rows less to give them a little more room. I find this helps the radishes all mature at the same time, so that I can cleanly crop out the entire bed at once. This allows me to free up that bed for something else, but also ensure the quality is consistent.

**Harvesting:** When we are bunching radishes for farmers markets, we bunch them in the field by eye. When packing for restaurants, we harvest them loose, tearing off the greens. If a bed of radishes is all mature at once, we will crop out the entire bed at the same time. Working our way down the bed, we put the bunches into a harvest bin that sits on the harvested part of the bed. We sometimes tear the greens off right after we've made the bunch, and throw them into another bin that goes into the compost. The other way is to thin harvest. During cooler months, the beds will not all mature at the same time. Usually the outside rows will be ready first. We harvest those first, and then over the next week will finish off the bed. When we're thin harvesting, we don't have room for a harvest bin because our walkways are too narrow. In this case, we leave the bunches in small piles in the walkways, then collect them into a bin once we've made our way through a bed.

**Crop Type:** Quick
**Seasons in production:** April through November
**CVR:** 4/5 (short DTM, high yield, long Harvest, popular)
**Planting Specs:** 5–7 rows; Jang roller F24; 0.7 oz per bed
**Varieties:** Easter Egg, French Breakfast, White Icicle, Raxe
**DTM:** 28 days
**Average yield per bed:** 70 bunches (35 pounds)
**Average gross profit per bed:** $175

Radishes

### Red Russian Kale

Red Russian Kale is a beautiful, robust and long-yielding salad green. It can be grown full size, much like any other kale, but I find it's best grown as a baby green for salad mixes or as a stand-alone salad green. It is a great crop to offer chefs because it can be available all season, so it's something that can be featured consistently on menus and it's not something you commonly find in a grocery store. Chefs like that. The shape of the leaves displays beautifully on a plate. It can be served very minimally with just a few leaves on a dish, but it also works well as a salad component and even slightly braised or wilted down. We are targeting a premium product, so we harvest it small: leaf and stem should be between three and four inches long.

**Production:** Red Russian is incredibly hardy and will withstand winter temperatures of 5°F. It's one of the first greens I direct seed in the greenhouses every late winter; my earliest seeding would be late February. One bed can be cut many times; I've counted as many as nine. You know when it's running near the end of its life: it starts to turn a dark purple and then the edges will go yellow. It still looks beautiful, but will start to get bitter. Unlike other kale, I find that Red Russian is far less prone to summer aphid pressure. If you are growing it in an area next to full-size kale, it will be more likely to attract aphids that are on your main kale. I approach this like most other pest problems, I'll just start planting Red Russian at another plot where there are no aphids. Because of this, I've been able to grow Red Russian all season without problems. After the first planting in late February or early March, I plant it every three weeks for the entire season.

**Harvesting:** We're harvesting in the same way as arugula: either with sharp serrated knives for hand harvesting or the Quick Cut Greens Harvester. We cut it at a young stage, and this way we get many cuts. During summer months, the crop is ready to harvest in 21 days from seeding. Red Russian yields fewer cuts in the summer than in the spring and fall.

**Crop Type:** Quick

**Seasons in production:** April through November

**CVR:** 5/5 (short DTM, high yield, high price, long season, popular)

Red Russian Kale

**Planting Specs:** Direct seed at 9 rows; Seed use: .6 oz; Seeder: Jang F24 roller

**Varieties:** Red Russian

**DTM:** 30 days

**Average yield per bed:** spring: 42 lbs per bed from 4–9 cuts; summer: 25 lbs from 3 cuts

**Average gross profit per bed:** $250 per bed

### Salad Mixes

Salad mixes are my biggest and most lucrative crop. They are a staple at farmers markets, but I move the most amount of volume to restaurants. Greens and mixes are hands down the most profitable crop you can grow on a small land base: they grow fast and can be cut multiple times. What's great about growing mixes is the fact that you don't have to have the exact same amount of greens going in the mixes each week; this offers a lot of flexibility for when you have an abundance or scarcity of any particular crop that may be a part of the mix. The expansion of my salad mix production is what made my farm more than double its gross profit from one year to the next, and it was the inclusion of restaurant clientele that would sometimes buy up to one hundred pounds per week that allowed me to move that kind of volume.

I offer three types of salad mixes: braising mix, spring mix and spicy mix. Each has its own peak season. Most of my customers like the fact that the greens are seasonal.

*Braising mix* is one of the first mixes that I offer during the season. It consists of tatsoi, red russian kale, arugula and mustard greens. It's called a braising mix because it can be braised (lightly cooked) to be part of soups and stir-frys. Because all of these greens are very cold tolerant, they flourish during the cooler months. Braising mix is available from the first week of April through mid-May. At this point, the weather is too hot to grow tatsoi, which is a critical part of the mix. After May, I can continue to grow all of the other components, and the arugula and mustard become Spicy Mix. A summer version of braising mix can include beet greens, mustard, arugula and Red Russian.

*Spicy mix* consists of mustard greens and arugula. It's available starting the same time as braising mix, but can be grown further into the season because the mustard and arugula aren't as quick to bolt as tatsoi.

*Spring mix* is my staple mix. It consists of a few types of lettuce and whichever seasonal greens I have at the time. In the spring time, it consists of lettuces, tatsoi, arugula, mustard and Red Russian kale. In the summer, it'll have more lettuce and less of the other greens because some of the cooler-weather greens will slow down production in the hot months. During summer, I grow baby beet greens to add to the mix that is 90% lettuces.

Each of the components is grown separately. Tatsoi, mustard greens and beet greens are used only for mixes, where Red Russian kale and arugula can be marketed separately.

**Production:** Success with spring mix requires having very strong lettuce production (outlined earlier in this part). Look at your lettuce production as the main part of the mix, and all of the other greens as secondary. If you have an abundance of arugula one week and are producing more than you can sell, then you can add more to your mixes. There are also times in the summer where my spring mix is 100% lettuce.

**Crop Type:** Quick

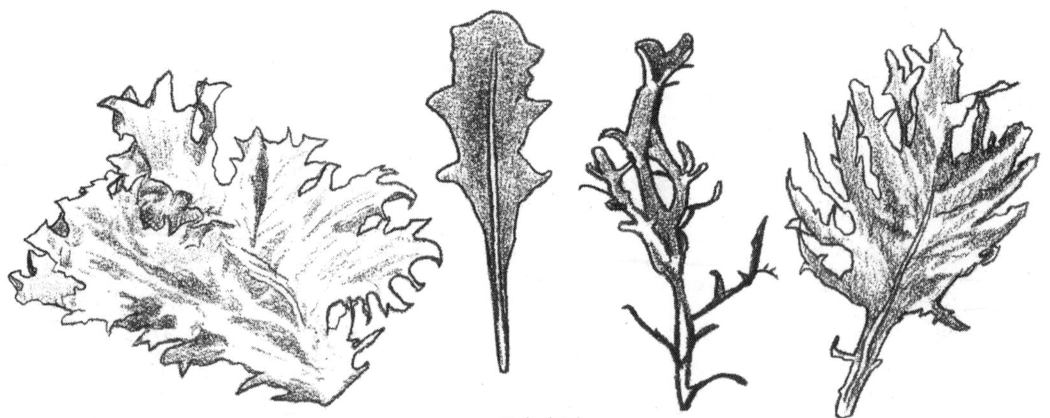

Salad Mix

**Seasons in production:** Mid-April through November
**CVR:** 5/5 (DTM, high yield, price, season, popular)

### Salad Turnips

In my experience, Hakurei or Tokyo turnips (as some chefs refer to them) are incredibly popular with high-end restaurants, less so with farmers markets. However, I do grow them for market on occasion; I find that customers there prefer a tennis-ball-size turnip with about three in a bunch, where the chefs prefer a small ping-pong-ball-size turnip in a bunch weighing around ½ pound with only a few inches of greens left on. When chefs display these turnips on a plate, much like other roots, they want an inch or less of green to be left on the turnip. Hakurei can be challenging to grow, as during the spring, cabbage root maggot can be a major issue, and during the summer hot and dry weather can cause poor germination.

For these reasons, I charge a premium price for this crop, but I have found ways to mitigate these problems as well. For the spring time, I keep all plantings covered in insect netting.

**Production:** There are two ways I've planted salad turnips in the past, and both work well. The first is to plant five rows in a bed with tight in row spacing, about two seeds every quarter inch. This means they're crowded in the row, but will push apart into the wider row spacing. I've used the Earthway seeder using the radish plate for this. Using this planting method, I've pulled more than 200 bunches from a bed. The problem is that the crowding can sometimes compromise the crop with fungus issues. The best way I have found is to plant nine rows, with a slightly wider in row spacing, about one seed every ¼ inch. For this I use the Jang seeder with the YYJ24 roller. This gives the crop the perfect amount of space and allows for a faster and more consistent harvest. Yield is a little less, but still around 100 bunches per bed.

**Harvesting:** We harvest salad turnips in two ways, depending on where they are going. For markets, we bunch them in the field just like radishes. For restaurants, we sell them

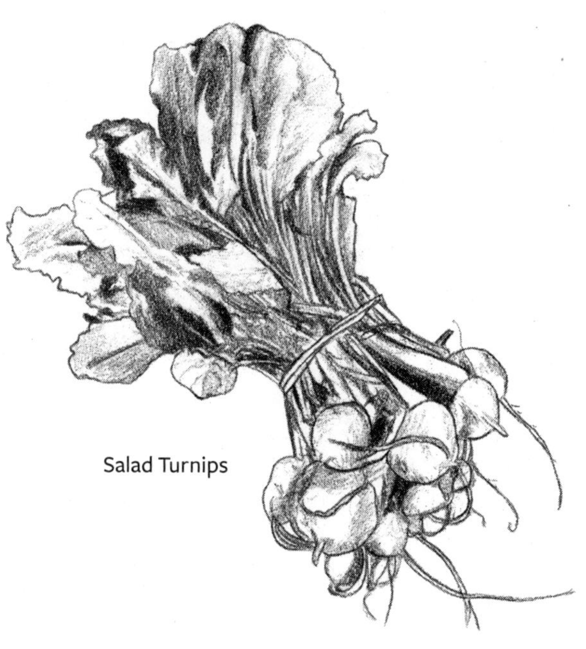

Salad Turnips

by the six-pound case. In this instance, we don't bunch them at all. As we harvest, we tear off the greens, leaving a few inches, then toss them in our bin. When they come back to home base, they are washed and packed into case lot boxes.

**Crop Type:** Quick
**Seasons in production:** May through October
**CVR:** 5/5
**Planting Specs:** 9 rows per bed; .17 oz seed; Jang seeder YYJ24
**Varieties:** Hakurei
**DTM:** 38 days from seed
**Average yield per bed:** 100 bunches at .5 lb each with tops removed
**Average gross profit per bed:** $300

## Scallions

Scallions, also referred to as spring onions or green onion, are a versatile crop that can be harvested almost all year. I sell small four-ounce bunches at the farmers market and in bulk to chefs. It's not a super popular crop, but is staple enough that everyone knows what it is and how to use it. What I really dislike about scallions is washing them: I find they're a lot of work based on what you can sell them for. When I sell them to restaurants, I find they're a little less work because they don't need to be bunched or as methodically cleaned as if they were on display at the market.

**Production:** A great thing about this crop is it's yield: one bed can grow up to $320 worth in about 70 days. I can harvest off one bed for three weeks. It's not something I sell a lot of, so I plant it only around four times a season. I start with an early crop from transplants planted into the high or low tunnels. After that, I direct seed it, just like any other salad crop. Planting it this way greatly reduces labor and increases yield.

**Harvesting:** We harvest scallions just like carrots. Use a fork to loosen the soil around them, then pull them and place them in a bin. Once I harvest a bin's worth, I will fill the bin with some water and stand the scallions upright so that the roots soak in water

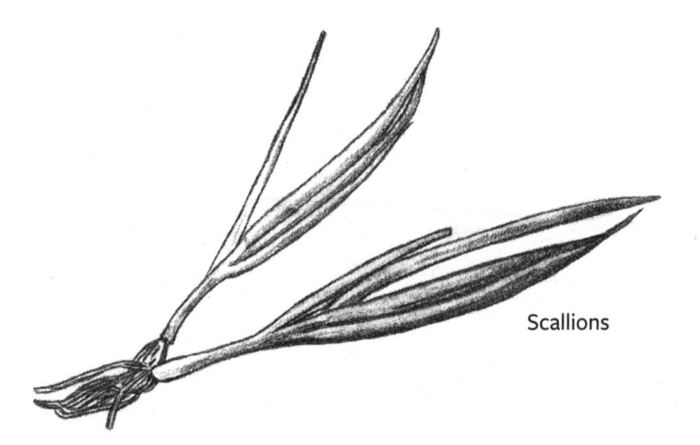
Scallions

for a few hours. This makes cleaning them a lot easier.

**Crop Type:** Steady

**Seasons in production:** CVR: 3/5 (high yield, long harvest, season)

**Planting Specs:** Transplanting: 6 rows, 4" apart; Direct seeding: 7 rows, ½ oz., Jang Roller F24

**Varieties:** Kincho

**DTM:** 70 days from seed

**Average yield per bed:** 40 lbs

**Average gross profit per bed:** $320

## Spinach

Spinach is a staple on my farm. I can grow a ton of it in a very small space. I sell a lot of spinach during the shoulder seasons and winters. I market two types of spinach: baby and broad leaf. The baby spinach is a premium product that mostly goes to restaurants to be used for salads in the spring, and I sell it in six-ounce bags at the market for $3. The broad leaf spinach is sold to both markets and restaurants as well, but is priced differently because it's used more for cooking and juicing. In this case, the price is considerably lower ($5–$7 per pound), depending on the season. Spinach is a profitable crop at both types and price points because the yields are much higher for the broad leaf crop.

**Production:** I plant spinach all seasons except in summer. My first big harvest is around mid-March, and that comes from an overwintered crop planted in early October. I will cut these beds until the first field sowing in the spring is ready. I plant the first field sowing around the third to fourth week of March. From here, I plant spinach weekly and bi-monthly until mid-June. By this time it's too hot for spinach to germinate well, so I take a break until mid-September. By mid- to late September, I will do two more plantings that will be a fall crop, then I finish the season by planting my overwintered crop during the first week of October. I plant five rows for a thick stand that can be harvested multiple times during the cool months. As the weather gets warmer, the crop starts to lose flavor

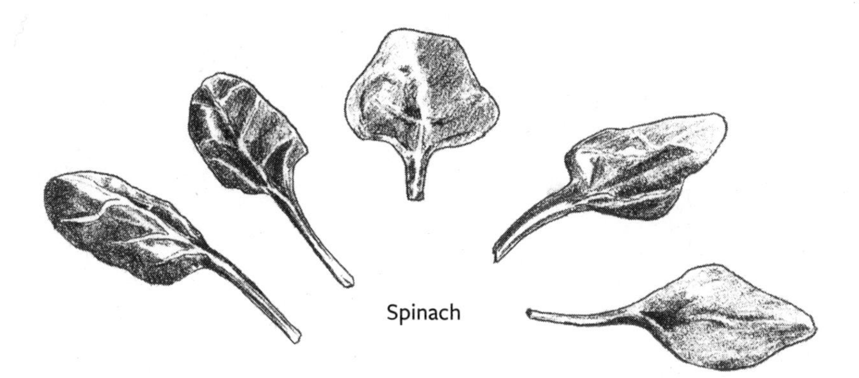

Spinach

and shape. Spinach is at its best during cool months, and even in winter it has a sweet flavor. For the first harvest of the overwintered crop in March, beds can be cut up to four times, and as the weather gets warmer, less and less. I know it's time to stop planting spinach when I get only one cut before it goes to seed.

**Harvesting:** We harvest spinach by hand and knife in the same way as most other greens. I have found that the Quick Cut Greens Harvester doesn't work well for spinach, as it misses the bottom leaves. I find I get more yield and a better product by hand harvesting.

**Crop Type:** Quick

**Seasons in production:** March through early June; Mid-September through December

**CVR:** 5/5 (DTM, high yield, season, price, popular)

**Planting Specs:** Direct seeded 5 rows; Earthway seeder, Chard plate; 1.9 oz per bed

**Varieties:** Space

**DTM:** 45 days

**Average yield per bed:** 35 pounds 2 to 4 cuts

**Average gross profit per bed:** $245

## Swiss Chard

Chard isn't an incredibly popular crop on my farm, but I like to grow it to have variety for chefs and on my market table. Chard plants have high yields and are easy to grow. I favor a variety called Bright Lights, which has a mix of colors. Chefs seem to like it, and I can sell ten pounds a week or more to them (in five-pound cases) and 10–20 bunches at the market. I usually plant only one bed of chard per season. I can cut it many times, and the plants never go to seed. I have also never had any major pest problems, except some slugs in the spring.

**Production:** Chard is planted to the same specs as kale. Chard can be grown almost all year long. It can be overwintered, but I find that it's not quite as resilient as kale.

**Harvesting:** Similarly to kale, tear off the bottom leaves, but it can also be cut clean off with a knife for a fast harvest. If you do this, chard plants takes a little longer to regenerate.

**Crop Type:** Steady

**Seasons in production:** May through October

**CVR:** 3/5 (high yield, high price, long season)

Swiss Chard

**Planting Specs:** Transplants: 3 rows, 10" centers
**Varieties:** Bright Lights
**DTM:** 65 days from seed
**Average yield per bed:** 65 pounds, multiple pickings
**Average gross profit per bed:** $325

### Tomato

Tomatoes are a staple summer crop for most farmers. The challenge to growing a marketable tomato is to grow different varieties from what everyone else is growing. My farm is hot and dry in summer, perfect tomato growing weather. Because of this, most farmers in my area grow a lot of tomatoes, and this means the market can be pretty saturated during these hot months. I noticed that, because there are so many organic farmers in my area, everyone was growing heirloom varieties, and very few were actually focusing on a generic salad tomato or slicer. I found a niche, growing hybrid saladettes and cherries for restaurants and farmers market. I do grow some heirloom beefsteak types, but they are all indeterminate varieties. My favorite varieties are Sun Golds and Sakura for cherries, Mountain Magic and Golden Rave for saladettes, Oxheart and Vintage Vine for large heirlooms and San Marzano for saucing.

**Production:** I grow only indeterminate varieties on my farm. An indeterminate variety will bear fruit over a longer period of time, where a determinate will set its fruit all at once. All my tomatoes are grown with a method we call Hard Pruning. The tomato is planted in dense spacing, only ten inches apart in the row. They are trellised with bailing twine and twisted up with twine as they grow. We prune every single sucker off the main stem, as well as every single branch beneath the last ripening set of fruit. Stripping off the foliage has a four-fold effect:

1. We're forcing the plant to focus its energy on ripening fruit and not growing foliage.
2. Removing foliage creates airflow near the bottom of the plant, which helps prevent disease and fungus problems.
3. We bring more light onto the soil as the sun recedes further into the southern sky during fall; this creates opportunities to interplant fall and winter crops amongst the tomatoes in September and October.
4. It makes cleanup at the end of the season fast and easy.

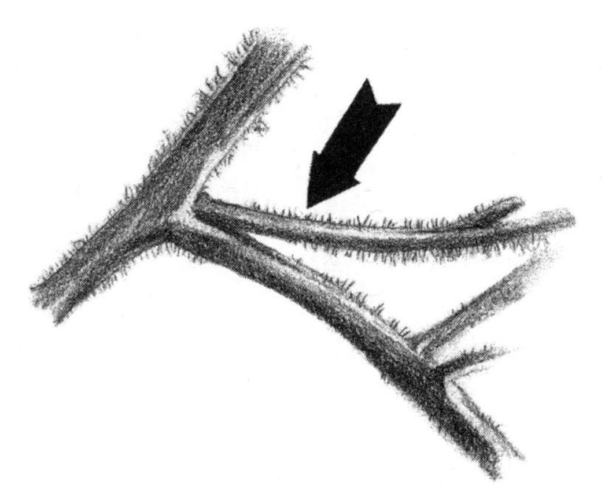

The arrow is pointing to a sucker branch. All of these need to be pruned off consistently to train the plant upwards.

See photo #53 in the photo insert section.

**Harvesting:** Picking tomatoes isn't very complicated, but it's important to harvest them into shallow bins so they don't get too stacked; too much weight on the fruit can cause them to bruise. We harvest into shallow bins, so there is only one layer of tomatoes. With cherries, you can have them a couple layers deep, but not with large, soft varieties like Oxheart or other heirlooms.

**Crop Type:** Steady

**Seasons in production:** July through October

**CVR:** 3/5 (high yield, price, popularity)

**Planting Specs:** Single row, 10 inches apart.

**Varieties:** Cherries: Sun Gold, Sakura; Saladette: Mountain Magic, Golden Rave; Large heirloom: Oxheart, Vintage Vine; Saucing: San Marzano

**DTM:** 145 days from seed

**Average yield per bed:** 180 pounds

**Average gross profit per bed:** $720

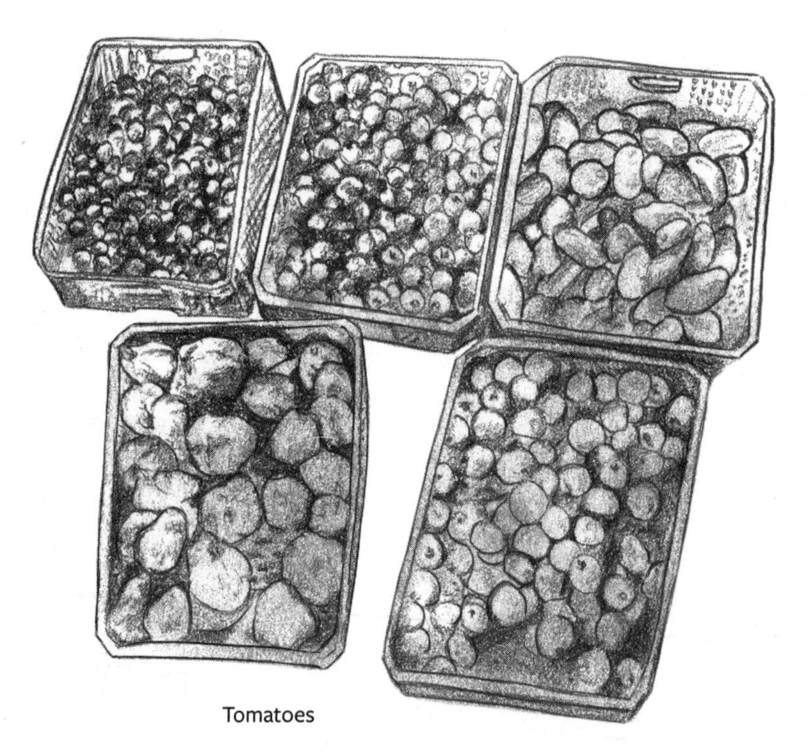

Tomatoes

# Parting Words

Looking back to starting my farm, there was never a time that I felt totally ready. In fact, I was so daunted by the whole idea that I would toss and turn throughout the nights. There were so many unknowns; what if this doesn't work? What if I fail? There were so many things that I didn't know how to do, and still so many questions unanswered. To be honest, not much has changed today.

There are always unknowns, and always more questions. Through experience you gain more confidence to execute tasks better and faster, but you never stop learning. Take comfort in knowing that there is always more to know, and don't ever let fear stop you from taking action or trying something new. You don't need to have all the answers in order to get started. You just need to put one foot in front of the other, and start moving. Figure out the details as you go.

Soon, you'll see that others will follow you, and they'll help push you along the way. Every step of the way after the first gets easier and more fun. This is exactly what has happened for me through embarking on my journey into farming, and it has also happened for many other urban farmers I have helped over the years. Our world is dying for good news, and urban farming is some of the most exciting news there is in the world of sustainable agriculture.

To build a better world all we have to do is look at problems as opportunities. Join me, and let's live a new definition of what it means to be a farmer: one who is at the root of the community by serving the needs of those in it. Farmers connect people to where their food comes from and empower others to start growing food themselves. Farmers can turn derelict areas into beautiful and productive farms that are flagship examples of local food resilience.

Let's take the best of what it used to mean to be a farmer—like the hard work ethic and self-determination—and leave behind the image of the grumpy, broke and destitute farmer, by being as profitable and successful as we can. A beautiful part is that the more productive your farm is, the more people will start to look, listen and ask questions. A profitable farm needs very little explanation. People will walk by and immediately understand it. This is very important if we want to disseminate the idea of urban farming around the world. Production needs no introduction. Get out there and get started. This revolution starts right in your own backyard, or someone else's!

# Acknowledgments

It's said that you are the sum of the closest five people you surround yourself with. I heard that right around the time I started farming, and it really inspired me to seek out mentors and absorb as much of their experience as possible.

Firstly, I'd like to thank my parents. My mother, Cheryl, for your love and support, and always believing in me. My father, Glen, for always teaching me the value of hard work, the virtue of integrity, and for always being there for me when I had questions about life and business.

My early mentors were so generous with their time. Your wisdom, knowledge and encouragement was critical to my success: Gwen Steele, Jon Alcock, Wolfe Wesle and Wally Satzewich. Thank you so much.

# Glossary

**Bi-Rotation:** describes an area where only two crop successions grow during one season. Farm plots that are furthest in the network will most often be BR because BR requires less work. Greenhouses are one exception. A BR area consists of a Primary Crop and a Secondary Crop. The Primary is a the Steady Crop that is in the ground the longest (for example, tomatoes). The Secondary Crop is usually a Quick Crop that either precedes or follows the Primary. For example, a Quick Crop like spinach can be planted and harvested in the same area as tomatoes before the tomatoes are even set into the ground.

**Board Technique:** a way to grow microgreens in soil on the ground using a sheet of plywood to germinate the seeds

**BR:** see Bi-Rotation

**Community Supported Agriculture:** a way for farmers to connect directly to eaters through a membership-based program. People buy a share (like stock) in the farm and receive a weekly box of seasonal vegetables determined by the farmer (simply what's available during that season). By paying up front for vegetables, customers are sharing in the risks and rewards of the farm. Today more CSA farmers allow for customization of the boxes through web-based ordering platforms.

**Crop Out:** to completely harvest a crop from the entire bed, leaving nothing behind

**Crop Value Rating:** used to determine the economic value of a crop for an urban farm. The smaller your land base, the higher you want the crops you grow to score on the CVR. Five main characteristics each represent one point in the score: (1) Shorter days to maturity; (2) High yield per linear foot; (3) Higher price per pound; (4) Long harvest period; (5) Popularity.

**CSA:** see Community Supported Agriculture

**Cut and Come Again:** loose-leaf greens that come back for repeated harvests. Any greens that can be cut multiple times (for example, arugula, lettuce, mustard and red Russian kale) are Cut and Come Again crops.

**CVR:** see Crop Value Rating

**DTM:** Days to Maturity, the time it takes a crop to grow from planting to harvest.

**Field Heat:** When vegetables have been harvested during warm weather, they will be warmed by the sun. To ensure longer shelf life and a consumable product, field heat must be removed from harvested crops quickly. Cold storage and cool water are typically the best way to do this.

**Field Micros:** microgreens that are grown outdoors in the soil, in my area between mid-May to mid-September.

**High Season:** the time of year where days are the longest and the farm is in its most productive state. On my farm, crops like arugula and radishes are mature in 21 days, and summer crops are picked almost daily. In the Northern hemisphere, high season is the months of mid-June through mid-September.

**Hi-Rotation:** areas of the farm that are in constant succession. Each bed in an HR area will be planted up to four times in a season. HR areas typically just grow Quick Crops, though beets and carrots in the high season are exceptions.

**HR:** see Hi-Rotation

**Plating:** a chef's term, meaning how a product is displayed on a plate.

**Post-Harvest:** describes all stages of a crop or product after harvest. This includes washing, sorting and bagging.

**Processing:** washing and sorting of vegetables after the harvest. See Post-Harvest.

**Primary Crop:** the main crop in a Bi-Rotation area (such as tomatoes, kale or pattypan squash). A Primary Crop will occupy that land for most of the season.

**Quick Crop:** crops that are mature within 60 days or less. Radishes, spinach, loose lettuce, and salad turnips are all Quick Crops. They are being planted and harvested nearly every week during the high season.

**Secondary Crop:** what precedes or follows a Primary Crop in a Bi-Rotation area. It is most often a Quick crop that (1) has a short enough DTM to be able to be mature before the Primary Crop goes in or (2) will be mature before the season end after the Primary Crop has been pulled out.

**Shoulder Season:** spring and fall. At this time of year temperatures are cooler, and much production on my farm is coming from greenhouses and poly low tunnels.

**Steady Crop:** has more days to maturity than a Quick Crops; it is harvested for a long period on a continual basis. Tomatoes, pattypan squash and kale are all Steady Crops.

**Thin Harvest:** to harvest crops that are mature and leave the rest. This is done most commonly with beets, and sometimes with radishes and turnips, especially early in the season.

# Endnotes

### Foreword

1. US Department of Agriculture Economic Research Service. "Farm Sector Profitability Expected To Weaken in 2015." Farm Sector Income & Finances, Highlights from the 2015 Farm Income Forecast, April 22, 2015. [online]. [cited July 6, 2015]. ers.usda.gov/topics/farm-economy/farm-sector-income-finances/highlights-from-the-2015-farm-income-forecast.aspx.

### Preface

1. World Wide Opportunities on Organic Farms website. [online]. [cited May 3, 2015]. wwoof.net/.

### Chapter 1

1. Francie Diep. "Lawns vs. crops in the continental U.S." Scienceline, July 3, 2011. [online]. [cited August 25, 2015]. scienceline.org/2011/07/lawns-vs-crops-in-the-continental-u-s/.
2. For the 85 million US households with a private lawn, an average lawn size is about 1/5 of an acre. Erin Chapman. "Lawn Size." *Grounds Maintenance.* [online]. [cited July 25, 2015]. grounds-mag.com/mag/grounds_maintenance_lawn_size/.

### Chapter 5

1. Johann Heinrich von Thunen. *The Isolated State.* Pergamon, 1966.

### Chapter 7

1. For example: Jean-Martin Fortier. *The Market Gardener: A Successful Grower's Handbook for Small-scale Organic Farming.* New Society Publishers, 2014; or Eliot Coleman. *The New Organic Grower: A Master's Manual of Tools and Techniques for the Home and Market Gardener,* 2nd ed. Chelsea Green, 1995.

### Chapter 9

1. F. John Reh. "Pareto's Principle—The 80–20 Rule." About.com. [online]. [cited April 30, 2015]. management.about.com/cs/generalmanagement/a/Pareto081202.htm.

### Chapter 14

1. The most common crowdfunding sites are: kickstarter.com, indigogo.com and crowdfunder.com. [online]. [cited May 7, 2015]. There are many more as well.

### Chapter 18

1. Both articles [online]. [cited June 24, 2015]: Gary Buiso. "Lead found in community gardens' soil, may affect produce." *New York Post,* March 26, 2014. nypost.com/2014/03/16/lead-found-in-community-gardens-soil-may-affect-produce/; Randy Shore. "Metal contamination found in Vancouver community garden, brownfield sites." *Vancouver Sun,* December 2, 2014. vancouversun.com/life/Metal+contamina-

tion+found+Vancouver+community+garden+brownfield+sites/10431761/story.html#__federated=1.

## Chapter 31

1. J. Rahkonen, J. Pietikäinen and H. Jokela. "The Effects of Flame Weeding on Soil Microbial Biomass." *Biological Agriculture & Horticulture: An International Journal for Sustainable Production Systems*, Volume 16#4 (1999), pp. 363–368. [online]. [cited July 13, 2015]. tandfonline.com/doi/abs/10.1080/01448765.1999.9755239#.VZXzQWC-Cp4.

# Index

**A**
accessibility to plots, 86–87, 121
agreements and leases, 99–100
all season production, 24–25
amendments, 112, 193
animals, 101–102
aphids, 103, 240
apprenticeships, 56–57
arugula, 220, 224–225
auxiliary greens, 225–226

**B**
baby dill, 231
banks, 76–77
base of operations
  coolers, 136–138
  drying table, 140–141
  location, 17, 135
  office, 136, 141–142
  organization of, 135–136
  portioning station, 141
  spinning machine, 139
  tool space, 138
  washing station, 139
base plan, 217–222
basil, 226–227
bed-preparation rake, 144
beds
  double, 189
  forming, 110–111
  interplanted, 190–191
  long, 190
  no-till beds, 192–194
  preparing, 113
  short, 188–189
  size, 119–120, 187–188
  turning over by tillage, 191–192
  unique beds, 188–191
  units of production, 188
beet greens, 225–226
beets, 22, 227–228
Bermuda grass, 95
bikes, 155–156
birds, 101–102
Bi-rotation (BR) area, 17, 18
board technique, 201–203
bok choy, 229–230
braising mix, 241–242
brand, 73–74
broadfork, 143–144
brokering, 49
budget and expenses, 67, 69
bunching herbs, 230–231
business, 37–38
buying clubs, 50

**C**
cabbage root maggot, 103
cameras, 102
Canada Thistle, 85, 96
carrots, 232–233
cash flow, 22–24
Cassavant, Bernard, xxi
cats, 101
chefs. *see* restaurants
cherry tomatoes, 22
Chomsky, Noam, xv
cilantro, 230, 231
city size and density, 16
Coleman, Eliot, xvii
community bonds, 75–76
Community Supported Agriculture (CSAs)
  benefit of, 34
  and brokering, 49
  customization, 47
  discussion of, 46–48
  and farm size, 30, 34, 46
  payment options, 47
  pickup times, 48
  referral program, 47
compost, xix, 97–98, 160
contaminated soil, 83–84, 96–97
CoolBot™ system, 137, 138
coolers, 136–138
costs. *see also* economics; finance options
  base of operations, 136
  coolers, 138
  farmers market equipment, 146

infrastructure, 90
irrigation, 127, 129, 132
keeping costs low, 39–40
nursery equipment, 148, 149
poly tunnels, 152
start-up costs, 4, 27, 75
tools, 143, 145
credit unions, 76–77
crop development, 219
crop planning
base plan, 217–222
Cut and Come Again Crops, 219–222
factors to consider, 211
outcome projection, 213–215
season availability, 214–215
succession planting, 218–222
crop profiles, 63, 64
crop selection
for ¼ acre, 33–34
for ⅒ acre, 33
categories of crops, 18
Crop Value Rating (CVR), 20–22
factors in, 19–25
specialty crops, 25
summer crops, 24–25
urban vs. rural, 8–9
winter crops, 24
Crop Value Rating (CVR), 20–22, 223
cropping out, 12
crowdfunding, 76
Cuban Special Period, 20
customer relations, 42, 43–44, 74
Cut and Come Again Crops, 12, 219–222
cutting your losses, 165

D
debris, 111
defensive location rotation, 103
demand, 16, 21–22, 35. *see also* markets
demographics, 16
dill, 231
direct seeding, 195–197
displays, 44
double reach beds, 187
drip irrigation, 123–124, 125–127
drying, 175
drying table, 140–141

E
economics. *see also* finance options
of ½ acre, 34–35
of ¼ acre or less, 29–31, 33–34
of ⅒ acre, 32–33
of Bi-rotation (BR) area, 18
and crop planning, 213–215
of Hi-rotation (HR) area, 12, 18
keeping costs low, 39–40
of labor, 59
of part-time farming, 31–32
quick breakdown, 11–12
education, 7–8
employees, 57. *see also* labor
end of suburbia, 3
equipment
coolers, 136–138
drying table, 140–141
for nurseries, 148, 149
portioning station, 141
sourcing, 29–30
spinning machine, 139
tools, 143–146
washing station, 139

F
farm size. *see* size
farm tasks, 160
farmer, 249
farmers markets
equipment for, 145–146
harvesting for, 169
as market stream, 30, 41–45
packing for, 179–180, 181
and work schedule, 31
farming. *see also* urban farming
obstacles, xii
profitability, xi
SPIN farming, xvii, xviii
tips for success, xiii
fencing, 85–86
fertility, 97–98
fertilizers, 98
Field Bindweed, 85, 96
field heat, 136, 170–171, 174
finance options, 75–77
first mover advantage, 16, 73
flagship plots, 86
flame weeding, 163–164
flea beetle, 103
flow through system, 125
food trends, 53
Footer, Diego, xi

G
garden layout, 119–120
grass
domestic vs. invasive, 107
invasive, 95–96
removal, 109–110
smothering, 108

greens
auxiliary greens, 225–226
harvesting, 171
processing, 173–175
removing from root crops, 175–176, 233

H
hand planting, 195
hard pruning, 247
hardpan, 97
harvest period, 21
harvest tally, 67
harvesting
arugula, 224–225
auxiliary greens, 225–226
balancing, 161
basil, 227
beets, 228
bok choy, 229–230
carrots, 233
cold storage, 171
efficient movement, 167–168
equipment, 168
equipment for, 145
greens, 171–172
harvest sheet, 169
herbs, 230
kale, 234
lettuce, 235–236
microgreens, 237
order of harvest, 170
pattypans, 238–239
protocols, 171–172
radishes, 239–240
Red Russian kale, 241
removing field heat, 170–171
root crops, 172
scallions, 244–245
spinach, 246
stages of, 168–171
structuring the week, 168–169
summer crops, 172
Swiss chard, 246
time of day, 170
tomatoes, 248
tracking, 171
turnips, 243–244
zucchini, 238–239
heat island effect, 4, 205
herbs, 230–231
Hi-rotation (HR) area, 11, 12, 17, 18
home base. *see* base of operations
hoop houses, 153
human pests, 102
hydroponic operations, 91

I
indoor vertical nursery, 146–150
infrastructure
base of operations. *see* base of operations
introduction to, 27
major purchases, 136
poly tunnels, 151–153
special growing areas, 146–150
time required, 29
tools, 143–146
insects, 102–103
interplanting, 190–191
invoicing and accounts, 70–71, 141–142
irrigation
approach to, 123–124
budget, 127, 129
budget and expenses, 132
drip systems, 123–124, 125–127
equipment, 132
greenhouse systems, 132
during harvest, 171–172
materials list, 127, 129
overhead systems, 123, 127–131
seasonal, 124–125
timers, 131
timing of, 124–125
types of, 123

J
Jang Seeder, 145

K
kale, 233–234

L
labor
apprenticeships, 56–57
economics, 59
employees, 57
productivity, 30, 58–59
starting small, 39–40
systems, 57–58
team building, 58
volunteers, 55–56
land
agreements and leases, 99–100
peri-urban land, 91
suburban land, 91
urban land, 89–91
land allocation data, 67, 68
land checklist, 82–87
landowners, 83, 169
landscape fabric, 122, 164
landscape rake, 144

lawn edger, 144
lawns, 3
lawns, converting
factors to consider, 107
large plots, 113
spring, 109–113
stages of, 109–113
summer or fall, 108–109
Lawton, Geoff, xi
layout, 119–120
leases, 100
lettuce, 221, 234–236
light, 85
localization, 7
location
accessibility, 86–87
choosing, 15–17, 81–87, 115–117
door to door, 82
fencing, 85–86
of home base, 17
invasive weed check, 85
landowner, 83
light, 85
logistical checklist, 82–87
low-hanging fruit, 82
mapping software, 81–82
satellite farms, 116–117
site history, 83–84
size, 84
soil testing, 84
span of sites, 84–85
and time, xx
visibility to public, 86
water access, 86
long harvest period, 21
lot size, 17

**M**

mammals, 101–102
mapping software, 81–82
markets. *see also* farmers markets
brokering, 49
buying clubs, 50
CSAs, 30, 34, 46–48
farmers markets, 30, 31, 41–45
importance of access, 4
and location selection, 16
restaurants, 30, 31, 32, 45–46
retail and other distributors, 49–50
specialized, 32
types of, xxi, 41
understanding, 35
media, 74
Memorandum of Understanding (MOU), 99–100
microclimates, 5
microfarm, 32–33
microgreens
field micros and board technique, 201–203
harvesting, 237
indoor and greenhouse growing, 199–201
processing, 173–175
production, 236–237
mulching, 121–122
multi-plot farms, 93–94
mustard greens, 221, 225–226

**N**

negotiating, 99
no-till beds, 192–194
nursery, 146, 197–198

**O**

office, 136, 141–142
one-off events, 53–54
operations
48-hour farm week, 161–162
cutting your losses, 165
farm tasks, 160
preventive weeding, 163–164
structuring the week, 162
weekly routines, 162–163
working smart, 159–165
orders past, 65
outcome projection, 213–215
overhead, 4, 39. *see also* costs
overhead irrigation, 123, 124, 127–131
overwintering, 209

**P**

Pareto Principle, 43
parking lots, 90
parsley, 231
part-time farming, 31–32
past orders, 65
pattypans, 191, 238–239
pedal power, 155–156
Pedal to Petal, xix
perfectionism, 160
perimeter, 121–122, 164
peri-urban land, 91
pests. *see also* weeds
birds and mammals, 101–102
defensive location rotation, 103
insects, 102–103
urban advantage, 5
pitchfork, 143
planting
direct seeding, 195–197
equipment for, 145
hand planting, 195

soil blocks, 197–198
sowing density, 196
transplanting, 195, 197
plantings, 62
plot map, 61
plot progress, 69
plot size, xx, 188
plugs, 198
Pollan, Michael, xi
poly tunnel greenhouses, 205
poly tunnels, 151–153
portioning, 179–185
portioning station, 141
power harrow, 194
pressure compensating, 126
preventive weeding, 163–164
price, 21, 43–44, 180
processing
bunched root crops, 175–176
defined, 173
greens and microgreens, 173–175
loose roots, 177
stages of, 173
production targets, 58–59
profitability of farming, xi
promotion, 44, 73–74, 81
pruning, 247
public speaking, 8

Q
Quackgrass, 95
quail, 102
questioning attitude, 160
Quick Crops, 11, 23–24
Quick Cut Greens Harvester, 145
quick tunnels, 153

R
raccoons, 101
radishes, 239–240
rakes, 144
rats, 102
record keeping
harvest, 171
invoicing and accounts, 70–71, 141–142
restaurants, 53
software, 61
spreadsheets, 61–72
voice memos, 70
Red Russian kale, 221, 240–241
rentals, 100
restaurant markets, 30, 31, 32
restaurants
approaching, 51–52
and brokering, 48–49
food trends, 53
harvesting for, 169
as market stream, 30, 31, 32
one-off events, 53–54
packing for, 180, 182–183
record keeping, 53
seasonal menus, 52–53
retail and other distributors, 49–50
rhizomes, 95, 111
rings of agriculture, 19
rodents, 102
role of urban farming, 7–9
rooftops, 90–91
root crops, 172, 175–176, 177, 209–210
rototiller, 144–145
rototilling, 110, 191–192
rubble, 97
rural to urban connection, 8–9

S
salad mixes, 241–243
salad turnips, 243–244
Salatin, Joel, xi, xii
satellite farms, 116–117
Savory, Allan, xi
scaling up, 30
scallions, 176–177, 244–245
season availability, 214–215
season extension
in cold climates, 207–209
hoop houses, 153
overwintering, 209
poly low tunnels, 151–153, 206–207
poly tunnel greenhouses, 205
quick tunnels, 153
storage crops, 209–210
in warm climates, 207
seasonal menus, 52–53
seed order and stock, 69
seeder, 145
self-promotion, 44, 73–74, 81
semi-diversified farm, 34–35
shade, 85, 119
sheet mulching, 121–122
Shepard, Mark, xi
shorter days to maturity (DTM), 20–21
single-plot farms, 94
size
½ acre, 34–35
¼ acre, 29–31, 33–34
1⁄10 acre, 31–33
of beds, 119–120, 187–188
microfarm, 32–33
minimum, 84
starting small, 39–40
SketchUp, 61
small farm broker, 49

social capital, 5–6
social connection, 5–6
software
invoicing and accounts, 70–71
mapping software, 81–82
record keeping, 61
spreadsheets, 61–70
voice memos, 70
soil
contamination, 83–84, 96–97
fertility, 97–98
hardpan, 97
invasive weeds and grasses, 95–96
rubble, 97
subsoil, 111–112
testing, 84
soil blocks, 197–198
soil-based rooftops, 90
soil-free rooftops, 91
sorting, 175, 177
sowing density, 196
special growing areas, 146–150
specialty crops, 25
spicy mix, 242
SPIN farming, xvii, xviii
spinach, 22, 220, 245–246
spinning, 174–175
spinning machine, 139
spoilage, 70
spreadsheets
budget and expenses, 67, 69
crop profiles, 63, 64
harvest tally, 67
land allocation data, 67, 68
orders past, 65
plantings, 62
plot progress, 69
seed order and stock, 69
spoilage, 70
weekly orders, 63–66
weekly sales totals, 67
yields, 62–63
spring mix, 242
stale seedbed technique, 163
start-up, 4, 29–35
start-up and overhead costs, 27
Steady Crops, 11, 24
sterilizing seed, 200
stirrup hoe, 144, 193
Stone, Curtis
backstory, xv–xvii, 37
barriers to farming, xviii–xx
early lessons, xx–xxi
first farm plot, xviii–xx
Green City Acres, xxi–xxii
reasons for farming, xvii–xviii
storage crops, 209–210
structuring the week, 162, 168–169
subsoil, 111–112
suburban land, 91
succession planting, 218–222
summer crops, 24–25, 172
sunlight, 119
Swiss chard, 246–247

T
tatsoi, 225–226
team building, 58
*The Isolated State*, 19
tilling, 110, 191–192
tilther, 193–194
time management, 159
timers, 131
tomatoes, 22, 190–191, 247–248
tool space, 138
tools
costs, 143
farmers market equipment, 145–146
harvesting equipment, 145
planting equipment, 145
types of, 143–146
tractor, 144–145, 194
transplanting, 195, 197
transportation, 155–156
trucks, 155
turnips, 243–244
Twitch grass, 95

U
units of production, 188
urban farming. *see also* farming
advantages of, 4–6
as a business, 37–38
reasons for, 3–4
role of, 7–9
urban land, 89–91

V
visibility to public, 86
voice memos, 70
von Thunen, Johann Heinrich, 19

W
walk behind tractor, 144–145, 194
washing, 173–174, 176, 177
washing station, 139
water, 5, 86. *see also* irrigation

weeds
  invasive, 85, 95–96
  preventive weeding, 163–164
  smothering, 160
  timing of, 164–165
  urban advantage, 5
weekly orders, 63–66
weekly routines, 162–163
weekly sales totals, 67
winter crops, 24
winter farming, 207–209
wood chips, 121–122, 164
working smart, 159–165
WWOOF (World Wide Opportunities on Organic Farms), xvi

**Y**

yield per linear foot, 21, 23
yields, 62–63

**Z**

zucchini, 238–239

# About the Author

Curtis Stone is the owner/operator of Green City Acres, a commercial urban farm based in Kelowna, BC. Farming less than half an acre on a collection of urban plots, Green City Acres grows vegetables for farmers markets, restaurants and retail outlets. After six successful seasons, Curtis has demonstrated that one can grow an extraordinary amount of food in a backyard and make a good living doing it. During his slower months, Curtis works as a public speaker, teacher and consultant, sharing his story to inspire a new generation of farmers.

For more about Curtis and his work, visit greencityacres.com.

Credit: Katie Huisman (katiehuisman.com)

If you have enjoyed *The Urban Farmer*, you might also enjoy other

## BOOKS TO BUILD A NEW SOCIETY

Our books provide positive solutions for people who
want to make a difference. We specialize in:

**Climate Change ◆ Conscious Community**
**Conservation & Ecology ◆ Cultural Critique**
**Education & Parenting ◆ Energy ◆ Food & Gardening**
**Health & Wellness ◆ Modern Homesteading & Farming**
**New Economies ◆ Progressive Leadership ◆ Resilience**
**Social Responsibility ◆ Sustainable Building & Design**

## New Society Publishers

### ENVIRONMENTAL BENEFITS STATEMENT

New Society Publishers has chosen to produce this book on recycled paper made with 100% post consumer waste, processed chlorine free, and old growth free.

For every 5,000 books printed, New Society saves the following resources:[1]

| | |
|---|---|
| 38 | Trees |
| 3,429 | Pounds of Solid Waste |
| 3,773 | Gallons of Water |
| 4,922 | Kilowatt Hours of Electricity |
| 6,234 | Pounds of Greenhouse Gases |
| 27 | Pounds of HAPs, VOCs, and AOX Combined |
| 9 | Cubic Yards of Landfill Space |

[1]Environmental benefits are calculated based on research done by the Environmental Defense Fund and other members of the Paper Task Force who study the environmental impacts of the paper industry.

*For a full list of NSP's titles, please call* 1-800-567-6772 *or check out our web site at:*

**www.newsociety.com**